D R E A M
television

cultural   television
planetary   network

for the world peace

emanuel dimas de melo pimenta

title: **Dream Television - Cultural Television Planetary Network for the World Peace**
author: Emanuel Dimas de Melo Pimenta
year: 2015
Television, humanitarian project

www.emanuelpimenta.net

ISBN-13: 978-1522724117
ISBN-10: 1522724117

cover: © Emanuel Dimas de Melo Pimenta

*to Durval de Noronha Goyos
and João Barba*

and to my daughter Laura

*many thanks to all people engaged in this
project, and in special to Vera Dantas*

You may say I'm a dreamer, but I'm not the only one. Hope someday you'll join us. And the world will live as one.

John Lennon

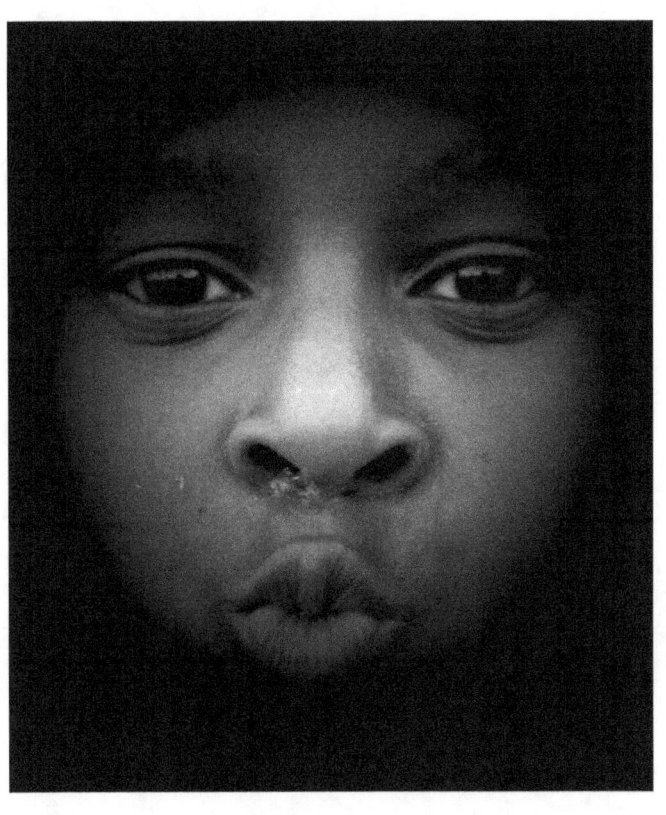

In the early 1990s I was in London having lunch with Peter Band, then a good friend. He told me that the OPEC Organization of the Petroleum Exporting Countries had funds for humanitarian projects.

He said he had a friend in that organization, they were looking for new ideas that would be important to the world, and Peter had talked about my work to that person, especially on the field of experimental architecture.

He challenged me to think about something.

I got a little lost. It seemed to be something too large. I asked him a few days to think.

At that time I was one of the coordinators of the first video art and electronic art festivals of the world, in Locarno, Switzerland, tpgether with René Berger, Rinaldo Bianda, Lorenzo Bianda, Nan June Paik, Edgar Morin, Madeleine Gobeil, Bill Viola and Pierre Restany among others.

I had participated, as one of the coordinators of the event, in Tim Berners-Lee's presentation of the World Wide Web - the launch of the www that would revolutionize the world.

My daughter Laura was born.

It was the moment the first Gulf War happened.

It was clear to me that Samuel Hundington was right when he said about a clash of civilizations - what was and continue being politically incorrect.

The world was developing in a mad rush to an ecological suicide, to a brutal social asymmetry, and of course to what would be showed by Islamic crazed fanatic terrorists years later.

It was not about revolted teens, imbalances of individual or of small groups. It was about different civilizational values.

There was no doubt - a new world war was intensified, underground and silently - and there was only one way out: communication, human formation, knowledge, discovery.

I reminded when Marshall McLuhan said, in the 1970s, that "all forms of violence are quests for identity. When you live out on the frontier, you have no identity. You are a nobody". But! The electronic universe placed everybody on the frontier, and especially the countries where literature has not create deeply roots.

An interesting project would be to establish a communication network of art and culture. Internet was not sufficiently developed for that yet.

The solution would be a TV network.

The initial goal would be the African Portuguese-speaking countries. Then gradually turn the idea into a planetary project.

A television network through which people knew more about their own cultural production, but also that of other countries, other cultures, other places. All under the sign of freedom, without ideologies or religions.

Freedom.

I sent the idea to Peter, who sent it to his friend. The answer came quickly: I should prepare a concrete proposal because the chances of getting a support from OPEC were real.

The support would come in terms of a long-term loan with virtually no interest.

That was in 1993. I started to design the project. Quickly, I called a dear friend, João Barba, great expert on television, who gave strong support especially regarding the technical data.

I and João Barba went to London to discuss the project and be able to develop it on more solid foundations and with better chances of getting the OPEC support.

Soon we realized that such project was much more important than we could imagine, it was important at a planetary level.

In 1996 I invited another dear friend for the project: Durval de Noronha Goyos, great lawyer, arbitrator of the World Trade Organization and one of the responsible for the legal design of Mercosul.

In this way, I, João Barba and Durval de Noronha started working as a team in order to improve the project.

On July 17 1996 the CPLP Community of Portuguese Language Countries was created, formed by Angola, Brazil, Cape Verde, Guinea-Bissau, Mozambique, Portugal, Sao Tome e Principe, and Timor Lorosae.

We decided to present our project as the first CPLP television.

We spend more four years in developments and negotiations until 2000, when Dulce Maria Pereira, then CPLP Executive Secretary embraced the idea.

But the terrorist attacks of September 11 in 2001 happened and, as George W. Bush had already announced, the world order had been changed, and the project was over.

All efforts to establish a network of culture and art was suddenly cut off.

About twenty years have passed since the start of the works for the implementation of this major project communication.

Now, I decided to publish the project - how it was presented at its time, because it could be useful to other similar projects.

The many graphics, tables and diagrams that constituted the original project are not included in this edition.

Descriptions country to country are very similar. This publication doesn't aim to bring a model for the implementation of a planetary television network, even because tecnhology in the last years the field knew a giant leap, not mentioning the emergence of the Internet.

This little book is dedicated to João Barba and Durval de Noronha Goyos, who have always been tireless enthusiasts of the project, always aware of its importance in humanitarian terms.

It is also dedicated to my daughter Laura, who was born just before the beginning of this adventure.

All thanks to everyone who participated in this dream.

Emanuel Dimas de Melo Pimenta

Locarno. 2015

CPLP

FICHA DE PROJECTO

Nome do Projecto: Rede de Televisão Cultural e Educativa dos Países de Língua Portuguesa

Número do Projecto:

Entidade Proponente: CPLP, NORONHA Advogados (São Paulo, Brasil), Duvídeo Profissionais da Imagem (Lisboa, Portugal).

Responsável pela Acção: Duvídeo Profissionais da Imagem e CPLP.

Data: Março de 2001

Local: Angola, Cabo Verde, Guiné-Bissau, Moçambique, São Tomé e Príncipe, e Timor Leste.

Duração: indefinida.

Justificação: Na Declaração dos Chefes de Estado e de Governo da CPLP sobre Educação, de 17 de Julho de 1998, foi afirmado o "compromisso de se organizar um conjunto de acções para aprofundar a cooperação comunitária na área educacional por meio de várias directrizes entre as quais ... o reforço do ensino médio, técnico profissionalizante, ... a promoção do ensino à distância nos diversos níveis e ... ampliar a disponibilidade de fornecimento de material didáctico de natureza variada e compatível com os sistemas de ensino de cada país". Excepção feita a Portugal e Brasil, o crescimento demográfico nos países acima referidos será fortemente relevante nos próximos vinte anos, segundo dados estabelecidos pelo Banco Mundial e pela Unesco. Torna-se, assim, urgente uma estratégia que atenda com eficácia às necessidades de formação quer no âmbito especificamente relacionado ao ensino sistematizado, quer à valorização das raízes culturais de cada região.

Países participantes: Angola, Brasil, Cabo Verde, Guiné-Bissau, Moçambique, Portugal, São Tomé e Príncipe, e Timor Loro Sae.

DREAM TELEVISION
Cultural Television Planetary Network for the World Peace

CPLP Cultural Television Network

The CPLP

The CPLP is an international organization created in 17 of July of 1996, with Headquarters in Lisbon and consisting of the following States Members: Angola, Brazil, Cape-Verde, Guinea-Bissau, Mozambique, Portugal and São Tome e Principe.

The CPLP has as its objective to promote the politic-diplomatic relationships between its Members, mainly referring to the reform of the United Nations and the World Bank, to the relationship EU/MERCOSUL and the accomplishment of the Europe-Africa Summit.

The CPLP is also dedicated to the co-operation, particularly in the economic, social, cultural, legal and technical-scientific matters; as well as to the projects of promotion and broadcasting of the Portuguese Language, nominated the improvement of the International Institute of the Portuguese Language and the creation of a Bibliographical Fund.

All its decisions are taken by consensus.

Executive Summary of the Project

General Background

Of Africa's estimated 778,500,000 people, about 65% live in rural areas. Only in Cape Verde, one of the most developed Sub-Saharan African countries, close to 30% live in poverty, and about 14% in absolute poverty. Poverty is predominately a rural phenomenon: 70% of the poor and 85% of the ultra-poor are in rural areas.

The incidence of poverty is highest among illiterate and female-headed households. Some 40% of all households in Cape Verde are headed by females and over a third are in poverty.

Some statistical information can give us a clearly framework of such a

dramatic reality:

The population in Africa was estimated in 778,500,000 people in the year of 1998. Following data from the World Bank and Unesco, up to the year of 2025 it is estimated a demographic growth launching the African population to 1,453,900,000 inhabitants, representing an increase of about 90% in only 25 years.

The energy consumption is another important indicator of the development tendencies in the African continent. The table bellow compares the consumption of some specific energy sources in Africa, Western Europe and the United States between 1986 and 1997. The information is in quadrillion Btu:

Petroleum

	1986	1990	1993	1995	1996	1997
Africa	3.67	4.17	4.56	4.83	4.91	5.11
Europe	26.54	27.42	28.37	29.61	29.97	30.33
USA	38.35	40.39	40.62	41.60	42.96	43.81

Hydroelectric power

	1986	1990	1993	1995	1996	1997
Africa	0.51	0.57	0.57	0.59	0.60	0.62
Europe	4.57	4.74	5.17	5.24	5.06	5.17
USA	6.85	6.38	6.71	7.18	7.87	7.80

Total Energy Consumption

	1986	1990	1993	1995	1996	1997
Africa	9.49	9.33	10.18	10.82	11.05	11.42
Europe	61.06	63.93	64.35	65.97	67.29	68.20
USA	88.82	97.02	100.79	104.92	108.43	109.14

In 1997, following the same data sources, the GNP per capita in the Sub-Saharan Countries was of about 510 dollars, against an average of 25,890 dollars in the Western Europe, the United States and Japan.

Also as a comparative table, between 1995 and 2000, the estimated infant mortality rates were:

Africa   86 per 1,000 live births

Asia   56 per 1,000 live births

Europe   12 per 1,000 live births

Northern America 7 per 1,000 live births

Considering some specific African countries speaking Portuguese, after Wistat's and United Nations' sources dated of 1990, the percentage of illiterates is:

Among 15-24 years old:

> ANGOLA: Women: N/A; Men: N/A
>
> CAPE-VERDE: Women: 13,6 %; Men: 10,1 %
>
> GUINEA-BISSAU: Women: 82,2 %; Men: 40,3 %
>
> MOZAMBIQUE: Women: 74,7; Men: 36,0
>
> SAO TOME & PRINCIPE: Women: 25,9 %; Men: 9,2 %

Among 25 years old and more:

> ANGOLA: Women: N/A; Men: N/A
>
> CAPE-VERDE: Women: 62,5 %; Men: 34,9 %
>
> GUINEA-BISSAU: Women: 95,7 %; Men: 77,8 %
>
> MOZAMBIQUE: Women: 93,8 %; Men: 65,8 %
>
> SAO TOME & PRINCIPE: Women: 36,9 %; Men: N/A

The percentage of illiterates in the whole Africa, between 1995 and 1999 is:

	1995	1998	1999
Men	35%	32%	31.1%
Women	52.9%	48.7%	47.4%

Communication is another interesting indicator of development. Based on information from Unesco dated of 1997/1998, comparing some African countries speaking Portuguese, Africa, the United Kingdom and the United States we have the follow picture (for each thousand people):

	1	2	3	4	5
Angola	-	0.3	-	4.7	34
Guine	-	-	0.5	7.3	42
Mozambique	0.8	-	0.4	3.4	38
UK	193	122	30.8	527.8	1433
US	362	165	53.9	639.9	2093
Africa	N/A	N/A	N/A	14.3	N/A

1 = computers; 2 = cell phones; 3 = fax; 4 = phones; 5 = radios

Africa is a continent submerged into dramatic problems. All these humanitarian problems have in the education and culture their most direct and objective root.

The African countries speaking Portuguese are among the African countries with the lowest level of development.

The whole situation is so dramatic that the implementation of the CPLP Cultural Television Network is more than urgent.

Project Objectives

The major objective of the project is to assist African countries – at the start countries speaking Portuguese – to develop the first African television channel oriented to cultural and educational themes.

This network will be named CPLP Cultural Television Network.

The development of such a project will create an irradiation pole from the countries with its implementation in the entire African continent, constituting an African network of research and culture, discovering and valorizing different cultural aspects and, consequently, reinforcing identity factors.

With a stronger cultural identity, it is expected to have lower levels of violence in social behavior as well as a relevant increase of the educational performance in all its sectors.

Systematic educational objectives are complementary to the general orientation of this project, which is cultural in its fundament.

A very large positive impact is also to be expected with respect to several sectors of the economy, greatly improving the overall quality of services.

The mission of the project is:

- rescue lost values and cultural information

- reinforce historical elements

- rescue old artworks

- give a new value to new artworks

- communicate practical and objective information on health

- communicate practical and objective information on agriculture

- communicate practical and objective information on economy

The CPLP Cultural Television Network will have in its philosophy, as a support to its mission, the follow fundaments:

- work with local people

- be focused on quality

- intercommunication

Three words can define the CPLP Cultural Television Network philosophy: **People, quality, intercommunication**.

Following these three key words of its philosophy, a single sentence also defines its mission: "A Continent of Culture".

Two main sectors characterize its global structure:

1. Cultural Sector
1.1. To develop programs on:

- music

- architecture (also vernacular)

- archaeology

- literature and poetry

- history

- tourism

- dance

- visual arts

- astronomy

- science in general

- gastronomy

- health

- anthropology

- local cultures

- others

1.2. The programs can be of all types, the most important is to be attentive to communication. All programs must reach the highest possible number of spectators, always with great quality.

2. Educational sector.

2.1. Systematic education in short practical courses.

2.2. Educational series with practical and objective information on:

- health care

- family

- agriculture

- managing

- services

- civil construction

As the CPLP Cultural Television Network must not be supported by the respective governments in the future, it will have a daily time of free commercial programs up to 12 hours. This is a good solution to pay its costs and support its development.

Mission Statement

To enhance the general levels of the cultural attainment of the population, reducing violence in general.

To improve the quality and relevance of the technical, vocational and professional skills.

To strength the institutional capacity to develop policies and implement programs in different areas.

The CPLP Cultural Television Network will not deal with:

- violence

- Nature destruction

- politics

- religion (in the sense to establish any judgements of value)

On the other hand, the CPLP Cultural Television Network will work on the improvement of the national, regional and local identity, reducing violence. It will defend human values and the preservation of Nature. It will not deal with Political Parties, and religion will always be took as a universal value, never giving to this or that religion in particular any special value.

The CPLP Cultural Television Network is a project for peace and its objective is not to compete with local television channels already installed, but yes to interchange with them, promoting a future global web dedicated to knowledge and development.

The first steps

The first objective of the CPLP Cultural Television Network is to install its studios in the follow countries:

- Sao Tome e Principe

- Cape Verde

- Mozambique

- Guinea

- Angola

- Timor

The next step will be to extend the CPLP Cultural Television Network also to South Africa.

Thus, these seven countries will structure the basic network of the entire project.

General description

The basic physical components of the project are:

1. The creation of companies, in each of the States of the CPLP and also in South Africa, oriented to the objectives above referred.

2. The State of each country – who will be the responsible and guarantee of the payment of the financing to OPEC – will also have a participation in the companies.

3. All strategies will be oriented to lowest costs in long term.

4. The whole project will be developed by phases.

5. The timetable of transmission will be defined case by case.

6. The project will include:

6.1. a complete plan for an equipment network

6.2. a complete plan for a software network

6.3. all procurements

6.4. equipment transportation and installation

6.5. all configurations

6.6. creation of the infographic system

6.7. technical formation of the personal

6.8. architecture and engineer projects for the building

6.9. construction of the building organized in phases

The project

Project Objectives

The main objective of the submitted project is to enable the countries of CPLP Community of Portuguese Language Countries to develop television channels oriented to cultural and educational topics with the following purposes:

- Dissemination of the Portuguese language

- Training of technical staff in the area of communication and tele-education

- Training and information to better participation of the peoples of CPLP in the overall development process

- Support the implementation project of the CPLP Portuguese Language Institute in Cape Verde

The development of such a project will create a process of irradiating pole of Portuguese-speaking countries, discovering and appreciating different cultural aspects and hence strengthening regional identity factors.

As result of a stronger identity, a reduction in the levels of violence in social behavior is expected as well as a substantial increase of education performances at all levels.

The areas covered by the project are:

- music

- architecture (also vernacular)

- archeology

- paleontology

- literature and poetry

- history

- tourism

- dance

- visual arts

- astronomy

- agriculture

- economy

- industry

- medicine

- science in general

- gastronomy

- health

- anthropology

- local cultures

- others

Some of the expected results of such a project are: increase health and education levels, reduce violence in general, improve the quality and relevance of technical, vocational and professionals skills, as well as to strengthen the institutional capacity to develop policies and implement programs in different areas.

The project also aims to serve as a basis for the mass production of health campaigns, particularly in the fight against AIDS, contagious infectious diseases, endemic or epidemic, through a comprehensive education process of regional populations.

Description

The general components of the project are:

1. The creation of a company oriented for the above purposes.

2. Each State will have a share in the respective company. This participation should be around 10% of the share capital.

3. All strategies will be targeted at achieving low long-term costs.

4. Another period of emission, equal to 3 hours per day, will be devoted specifically to education programs.

5. The project will involve:

5.1. the full development of network equipment plan

5.2. Complete drafting software network plan

5.3. all procurements

5.4. transportation and installation of equipment

5.5. Settings

5.6. development of the entire infographic system

5.7. Technical staff training

5.8. architecture and engineering designs for building

5.9. construction

Designation of the infrastructures:

1. Studio - 800 m2 using the architectural concept of deprogrammability.

2. Studio integrating a building with a total of up to 2,000 m2 of built area.

3. post-production equipment - non-linear post production, two stations ES7.

4. ENG - 3 DVCAM Camcorders

5. Régie 4 DXC-D30PK1 Digital Cameras - for cool lighting installation.

6. Flexicart.

7. CG Suite.

8. Sound Room.

9. MUST SYSTEM: Automatic Emmission Managing System, with graphic information for non-active emissions (This system has been used by several French thematic channels)

Obs. Duvideo / TGT Interactive are the exclusive representatives of this System for Portugal and African countries.

Other technical data - the installation is characterized by:

1. Computer network with level 5 UTP cables.

2. telephone system.

3. Specific furniture.

Discrimination of the technical data:

Studio

1. Cold lighting system.

2. 4 camera channels comprise by:

2 DXC-D30PK Sony Digital Cameras

4 J18 Canon Zoom Lens

4 CA-537P Sony Camera Adaptors

4 DXF-50 Viewfinder Sony Studio's

4 CCU M3P-Sony Camera Control Unities

4 RM-7P Sony remote CCU contols

4 50M Sony Multicore Camera Sets

4 Canon Focus and Zoom control sets

4 Headphones Intercom

4 Special tripods

1 charriot flat and curved

1 Teleprompter with 2 reading systems (monitors for individual speakers)

1 lighting control desk with memory - ARRI different microphones with different directional characteristics

Audio components

Video Components

General Intercom System

Video Régie:

1 1. Recorder / Player Betacam equipment

2. 1 digital video mixer with 12 tracks

3. 1 DVE equipment of digital effects for 3 channels

4. 1 TBC Frame Synchronizer

5. 2 Oscyloscopy and Vectorscopy for camera control

6. 2 Synchronizer Generators with changeover

7. 1 Intercommunication system with 4 places

8. 1 Audio and Video Matrix with remote controls

9. Video distributers

Audio Régie:

1. 1 audio mixer with 24 tracks

2. 2 audio monitors

3. 1 cd player

4. 1 cassette deck

5. 1 DAT

6. 1 Mini-Disc

7. 1 Level Detector for Audio Stereo

8. Stereo Audio Distributors

9. Audio compressors

10. Audio effects

11. Microphones

Video Post-Production:

1. 3 Sony ES-7 DVCAM / Avid non-linear hybrid edition stations

2. 1 Conventional Betacam Edition Suite 2: 1

3. Multitrack audio (8) on hard disc, Sound Scape type

3 ENG SETS:

1. Sony DSR-300 DVCAM Camcorder Digital Compact PK

2. Report projectors

3. Multiple support materials (batteries etc.)

Any material described above is supported by the peripheral functional devices.

BUILDING

The land will be donated by each respective state.

- Total construction area up to 2,000 square meters

- Land area: if possible from 15,000 to 20,000 square meters

General time limits

Architecture and engineering designs: 6 months*

Construction: 6 to 10 months*

Installation of equipment: 3 months after delivery of the building**

Preparation of the first operators: 2 months

Not considering any bureaucratic delays of local authorities and depending on specific construction characteristics.

** The start of operations will take place before the delivery of the building, partially and provisionally, in another location, in order to accelerate the schedule*

COSTS

The total cost of each unit (turn-key) is USD.11.600.000 (eleven million and six hundred thousand US dollars), paid in three years, funded by OPEC for a period of seventeen (17) years with a maturity grace period of five (5) years.

The OPEC funding covers the following points:

- architecture design and engineering

- Construction and monitoring of works

- Operating period of five years

- Purchase of equipment

- installation

- Settings

- Staff training

- And other items provided in the project description (see below)

Thus, the beginning of the return of financing by OPEC will take place five years after the start of the financing. The interest rate applied by OPEC will be extremely low.

Full implementation of the project estimate .... 1 year

Estimate for the break-even point ....5 years

Operating funding period ....5 years

The project (turn-key) for a cultural television station in Cape Verde means:

- Preparation of equipment network plan

- Full procurement

- Transportation and installation of equipment

- Adjustments and equipment configurations

- Architecture and engineering designs

- Construction of the building

- Monitoring of works

- Preparing the infographic system

- Staff training

- Supervision of the operation and management

FEASIBILITY

For the hiring of a system similar to other television producer for a period of transmission of 6 hours per day, the cost of market turns around USD.500,000 * per month. That means a total of about USD.6,000,000 per year. Over 16 years, discounted the first year of implementation, the cost of hiring another channel for the same six hours a day will be therefore about USD.100,000,000 - much higher than the value of the presented project.

These costs are based on current prices in Portugal, such as the program *Hora Viva - Segurança em Directo*, broadcast daily, excluding weekends, from 7 am to 10 am.

Another example, the cost of the CNL Channel News Lisbon in the first year of installation was about USD.10,000,000, and worse - both the equipment and facilities were rented and not owned by the broadcaster.

* Values taken in the Portuguese market, which is about 30% below the European average market.

The first ten months will be dedicated to:

- Design of equipment and software

structure

- Preparation of architectural designs and engineering

- Procurement and purchase of all equipment

- Construction and supervision of works

- Installation and equipment configuration

- Start-up

Thus, the first year will be mainly devoted to project implementation. However, in order to accelerate the schedule, some of the operations of the new television station could temporarily start before, in another building.

Estimates indicate that from five years of steady operation, the cultural and educational TV channel will export a large amount of services and videos to Africa, Europe and South America.

The economic feasibility of the project will also take place with the marketing of various products:

1. The sale and production of cultural programs for different institutions such as:

    - BBC

    - Discovery

    etc.

Worldwide, commercial networks oriented to cultural programs have been created. Several television stations in several countries regularly buy program producers scattered throughout the world. Not only this, several international organizations such as the World Bank and the United Nations among many others, require public education programs such as awareness programs and education as endemic or epidemic nature of diseases such as AIDS, malaria, tuberculosis etc. Or, programs that focus on new farming techniques etc.

These global institutions have been making great efforts to develop television programs with humanitarian objectives.

The price for each institutional campaign, with an average production of 10 (ten) films, is about USD.15,000 and the capacity of the Cultural and Educational CPLP Channel, within six months to a year, will be of two campaigns per month.

2. The sale and production of non-cultural programs to other television stations such as:

- Series

- Talk shows

etc.

With a single sale of a daily 3-hour program, for five days a week, for a price equivalent to that of Portugal - and therefore down by about 30% of the European average, and with the advantage of not having high transportation cost - the income of project for the cultural and educational television would be twice the entire investment.

The price for this kind of television programs is:

Talk shows USD.12,500 per program

Talk shows USD.4,000 per program

Interviews USD.4,000 per program

Production capacity, within six months to a year, of these programs at the cultural and educational channel will be:

Pop shows: 6 per month

Talk shows: 20 per month

Interviews: 20 per month

The price of soap operas is much higher. The price of each one of them, complete with about 120 chapters, is around USD.2,500,000 and the capacity of the Cultural and Educational Channel is to make one series a year.

3. It is open the possibility of commercial use of transmission time beyond six hours specifically reserved for cultural channel and education.

Conventional daytime television is 18 hours, of which only 6 would be not commercial. They would therefore be 12 hours of non-cultural spaces that could be freely traded.

Even the period of 6 hours - divided between three hours dedicated to culture and 3 hours to education - has a great potential to attract sponsorship.

Each commercial transmission time can be up to 12 minutes of advertising. The price for each 30 seconds of advertising is:

Prime time from 18:00 to 23:00 approximately USD.400

Regular time about USD.200

Concentrating the business hours in the prime time, with four hours of transmission in this period, we would have:

Educational television: from 7:00 to

9:00 and from 15:00 to 16:00

Cultural television: 16:00 to 19:00

Commercial television from 9:00 to 15:00 and from 19:00 to 1:00

So, the commercial period would comprise six hours during regular hours, 4 hours in prime time and 2 hours at regular time.

Being 12 minutes per hour of advertising, the commercial period could comprehend:

Prime time: 48 minutes, 24 films = USD.9,600 per day

Regular hours: 96 minutes 48 films = USD.9,600 per day

Total .... USD.19,200 per day or USD.576,000 per month

4. The lease of studios for television crews of other countries.

The lease period of a television studio is 12 hours.

Each period has a price of about $ USD.4,000. After the first year, it will be possible to lease the studio up to 12 hours every 3 days period. Thus, the rental per month in this period could generate revenues of USD.120,000 per month.

5. Support services to other television channels in all areas, including series, soap operas, miniseries, docdramas etc.

Each team outside its country has a cost of about USD.1,000 per day (two people), in addition to travel costs, room and meals. In the first months can be arranged up to 5 teams to work abroad. The capacity of each team is 20 days per month. Thus, the support to other television networks, if taken in its potential, may represent up to 100 days per month of work, meaning about USD.100,000 of monthly income.

6. The holding of symposia, seminars, video conferencing and meetings of different modalities.

The headquarters building of cultural and educational television can have an medium auditorium with full conditions to host seminars, conferences and meetings of various natures. The holding of such seminars, videoconferences etc., is not only important for increasing revenues, but it will also attract to Cape Verde experts from different areas, being an important element in the development of other projects and tourism.

The meetings, workshops and seminars join, on average, about 300 individuals per event. The price - except for food, stay and transportation - of each participant is about USD.50 per day. There will be a capacity for up to four of such events per month, which could represent a turnover of about USD.60,000 per month.

The project could also turn possible:

- Vocational training courses in various areas

- Support for high-tech multimedia center

- Support for multimedia cultural center

But it is important to note that although this new cultural and educational television channel will surely be a great commercial success, this is not its primary goal.

In such way, the feasibility of the project should be considered not only related to its financial return, but mainly to the urgent need of a project of this nature.

PROJECT REPORT

OF

ANGOLA CULTURAL TELEVISION NET-WORK

AT

ANGOLA

AFRICA

PROMOTERS

REPUBLIC OF ANGOLA

AFRICA

ANGOLA TELEVISION NETWORK

INTRODUCTION

CHAPTER - 1

INTRODUCTION

1.1   Description of the organisation through which this project is being taken up.

The CPLP is an international organisation, created in 17 of July of 1996, with Headquarters in Lisbon and consisting of the following State Members: Angola, Brazil, Cape-Verde, Guinea-Bissau, Mozambique, Portugal, Sao Tome e Principe and East Timor (Timor Lorosae).

The CPLP has as its objective the politic-diplomatic relationships between its Members, mainly referring to the United Nations and the World Bank, the relationship EU/MERCOSUL and the accomplishment of the Europe-Africa Summit.

The CPLP is also dedicated to the co-operation, particularly in the economic, social, cultural, legal and technical-scientific matters; as well as to the projects of promotion and broadcasting of the Por-

tuguese Language, nominated to the improvement of the International Institute of the Portuguese Language and the creation of a Bibliographical Fund.

All its decisions are taken by consensus.

2.2 Description of the main reasons for setting up an African Cultural Television Network.

Africa is a continent submerged into dramatic problems. All these humanitarian problems have in the education and culture their most direct and objective root.

The African countries speaking Portuguese are among the African countries with the lowest level of development.

The whole situation is so dramatic that the implementation of the CPLP Cultural Television Network is more than urgent.

It is hoped that with this Television Network the people shall have better understanding of various diseases and other problems of life. e.g. AIDS is prevalent disease in Africa. But as the people are illiterate the information about the disease cannot be spread by any other method. In such cases the Television Network becomes a very effective audio-visual means of communication.

The project also intends to be a base of a massive production of health campaigns, specially to reinforce the combat against the AIDS, the infect-contagious diseases- endemic or epidemic-through a process of global education of the local and regional populations.

The CPLP Cultural Television Network is a project for peace and its objective is not to compete with local television channels already installed, but to interchange with them, promoting a future global web dedicated to knowledge and development.

1.03 Description of the main objects of the project.

The major objective of the project is to assist the CPLP countries to develop the first African television network channel oriented to cultural and educational themes with the follow orientation:

-       diffusion of the Portuguese language
-       education in all areas
-       reinforcement of the local and regional cultures
-       formation of technical personal on communication and tele-education

-       formation and information for a better participation of the CPLP nations in the process of global development

-       to support the implantation of the Institute of Portuguese Language

The development of such a project will create an irradiation pole from the countries with the project implementation to the whole African continent, constituting an African network of research and culture, discovering and valorising different cultural aspects and, consequently, reinforcing identity factors.

With the stronger cultural identity, it is expected to have lower levels of violence in social behaviour as well as a relevant increase of the educational performance in all its sectors.

Systematic educational objectives are complementary to the general orientation of this project, which is cultural in its fundament.
A very large positive impact is also to be expected with respect to several sectors of the economy, greatly improving the overall quality of services.

1.4   Description of the mission status of the project.

The mission of the project is:

-       to rescue lost values and cultural information

- to reinforce historical elements
- to rescue old artworks
- to give a new value to new artworks
- to communicate practical and objective information on health
- to communicate practical and objective information on agriculture
- to communicate practical and objective information on economy

1.05 Description of the philosophy of the project.

The CPLP Cultural Television Network will have in its philosophy, as a support to its mission, the follow fundaments:

- to work with local people
- be focused on quality
- intercommunication

Three words can define the CPLP Cultural Television Network philosophy:

People, quality and intercommunication

Following these three key words of its philosophy, a single sentence also defines its mission: "A Continent of Culture".

Two main sectors characterise its global structure:
1. Cultural Sector

1.1.   To develop programmes on:
- Music
- Architecture (also vernacular)
- Archaeology
- Literature and poetry
- History
- Tourism
- Dance
- Plastic arts
- Astronomy
- Science in general
- gastronomy
- health
- anthropology
- local cultures
- others

1.2   The programmes can be of all types, the most important is to be attentive to communication. All programmes must reach the highest possible number of spectators.

2. Educational sector.
2.1   Systematic education in short practical courses.
2.2   Educational series with practical and objective information on:
- health care
- family
- agriculture
- managing

- services
- civil construction

As the CPLP Cultural Television Network must not be supported by the respective governments in the future, it will have a daily time of free commercial programmes up to 12 hours. This is a good solution to pay its costs and support its development.

1.06 Description the magnitude of the project.

The project proposes a Television transmission time of eight hours per day in the first year. It will increase by two hours per day in the second and subsequent years.

The project also envisages training of local people in all the related areas of Television. It is proposed to develop a local pool of talent which can be utilised not only in this particular country but also in other countries of Africa.

The total cost of the project of the Angola Cultural Television Network is US$.26.500.000. Out of this it is proposed to obtain a soft loan of US$.22.000.000. The remaining amount of US$.4.500.000 shall be contributed by the Government of Angola in terrain for the installation of the respective building, infrastructures and logistic support.

For the purpose of the Angola Cultural Television Network a company shall be formed with the participation of the Government. The Contribution of the Government in the form of share capital shall be US$.4.500.000.

The term loan shall be on soft terms. The total period of the loan shall be 23 years. Out of which five years shall be a period of moratorium. The entire loan shall be repaid in the remaining 17 years, in 17 equal yearly instalments.

The Government of Angola will give the sovereign guarantee for the referred loan.

ANGOLA TELEVISION NETWORK

THE COUNTRY

CHAPTER - 2

THE COUNTRY

2.01 Description the level of development of Angola.

Angola is an economy in disarray because of more than 20 years of nearly continuous warfare. Despite its abundant natural resources, output per capita is among the world's lowest. Subsistence agriculture provides the main livelihood for 85% of the population. Oil production and the supporting activities are vital to the economy, contributing about 50% to GDP.

Notwithstanding the signing of a peace accord in November 1994, sporadic violence continues, millions of land mines remain, and many farmers are reluctant to return to their fields. As a result, much of the country's food must still be imported. To take advantage of its rich resources-gold, diamonds, extensive forests. Atlantic fisheries, arable land, and oil deposits-Angola will need to implement the peace agreement and reform government policies. Despite the high inflation and political difficulties, total output grew an estimated 9% in 1996, largely due to increased oil production and higher oil prices.

OFFICIAL NAME: Republic of Angola
CAPITAL: Luanda
SYSTEM OF GOVERNMENT: Multiparty Republic
AREA: 1,246 700 Sq Km (481,354 Sq Mi)
ESTIMATED 2000 POPULATION: 12,897,000

LOCATION & GEOGRAPHY: Angola is located on the western coast of South Africa. It is bound by Namibia to the south Zambia to the east, Democratic Republic of the Congo (Zaire) to the north and north-east, and the Atlantic Ocean to the west. A separate province of Cabinda is enclosed by the Congo. A sparsely watered coastal plain extends along the coast and rises towards the interior which contains irregular terraces that form sub- plateaux. The central plateau accounts for around 66% of the land area and has numerous rivers which run into basins of the Congo and Zambezi Rivers that in turn flow to the Atlantic Ocean. The north-western region of the central plateau and the enclave of Cabinda are covered by equatorial jungles while the southern regions and coastal plain are semi-arid. The Namib Desert occupies the coastal plain above Mocamedes. Major Cities (pop. Est); Luanda 1.000.000, Huambo 203,000, Beneguela 155,000, Lobito 150.000 (1983). Land use; forested 42% pastures 23%, agricultural-cultivated 3%, other 32% (1993).

CLIMATE: Angola has a tropical climate with regional variations from a moderate tropical climate to a desert climate depending on the location. The prevailing winds are predominantly from the west, south-west and south. Average temperature ranges in Luanda are from 18 to 23 degrees Celsius (64 to 73 degrees Fahrenheit) in August to 24 to 30 degrees Celsius (75 to 86 degrees Fahrenheit) in March.

PEOPLE: Almost, the entire population is Bantu origin with various and numerous tribal groups. The principal ethnic majority are the Ovimbundu who alone represent 37% of the population while the Mbundu, Lunda, Chokwe, Nganguela, Ovambo, Herero, Kangala, Humbe, Luvale, Bunda, Luchazi, Kwandare and Cuanhama account for 38% of the population. The Ovambo and Herero are nomadic cattle herders that regularly migrate across the Angola-Namibia border.

DEMOGRAPHIC/VITAL STATISTICS: Density; 8 persons per sq km (21 persons per sq mi) (1991). Urban-Rural; 26.8% urban, 73.2% rural (1988). Sex Distribution; 51.1% male, 48.9% female (1990). Life Expectancy at Birth; 42.9 years male, 46.1 years female (1990). Age Breakdown; 42% under 15, 28% 15 to 29, 17% 30 to 44, 10% 45 to 59, 3% 60 and over (1990). Birth Rate; 47.2 per 1,000 (1990). Death Rate; 20.2 per 1,000 (1990). Increase Rate; 27.0 per 1,000 (1990). Infant Mortality Rate; 137.0 per 1,000 live births (1990).

RELIGIONS: Mostly Christians with around 70% of the population Roman Catholic while 20% are Protestant. Although, a large number still adhere to traditional tribal beliefs.

LANGUAGES: The official language is Portuguese, although the national language is Bantu with the number of dialects spoken as varied as the Bantu sub-tribes.

EDUCATION: Aged 25 or over and having attained: N/A. Literacy; literate population aged 15 or over 1,196,000 or 28% (1980).

CURRENCY: The official currency is the New Kwanza (NKz) divided into 100 Iwei.

ECONOMY: Gross National Product: USD $6,010,000,000 (1989). Public Debt; USD $7,727,000,000 (1993). Imports; USD $1,347,000,000(1997).Exports;USD$3,427,000,000 (1991). Tourism Receipts; USD $20,000,000 (1993). Balance of Trade; USD $1,565,000,000 (1994). Economically Active Population; 4,166,000 or 40.3% of total population (1991). Unemployed; N/A.

MAIN TRADING PARTNERS: Its main trading partners are Portugal, Cuba, Germany, the US, the UK, Canada, Japan and the former USSR.

MAIN PRIMARY PRODUCTS: Asphalt, Bananas, Cassava, Citrus Fruits, Coffee, Copper, Cotton, Diamonds, Fish, Gypsum, Iron Ore, Limestone, Maize, Manganese, Oil, Palm Oil, Phosphates, Salt, Sisal,

Sugar Beets, Sweet Potatoes, Timber.

MAJOR INDUSTRIES: Agriculture, Cement, Chemicals, Fishing, Food Processing, Sisal, Timber, Tobacco.

MAIN EXPORTS: Coffee, Crude Oil, Diamonds, Fish, Maize, Oil, Petroleum Products, Sisal, Timber, Tobacco.

TRANSPORT: Railroads; route length 2.789 km (1,733 mi) (1998), passenger-km 326.000.000 (203.000.000 passenger-mi) (1998), cargo ton-km 1.720.000.000 (1.178.000.000 short ton-mi) (1988). Roads; length 73,830 km, (45.876 mi) (1986). Vehicles; cars 122.403 (1989), trucks and buses 44,000 (1989). Merchant Marine; vessels 111 (1990), dead weight tonnage 122.403 (1990). Air Transport; passenger- km 975.000.000 (606.000.000 passenger-mi) (1985), cargo ton-km 33.900.000 (23.218.000 short ton-mi) (1985).

COMMUNICATIONS: Daily Newspapers; total of 4 with a total circulation of 500.000 (1994). Radio; receivers 450.000 (1994). Television; receivers 50.000 (1994) Telephones; units 53.300 (1993).

MILITARY: 82.000 (1995) total active duty personnel with 91.5% army, 1.8% navy and 6.7% air force while military expenditure accounts for 23.9% (1986) of the Gross National Product (GNP).

ANGOLA TELEVISION NETWORK

THE PROJECT

CHAPTER - 3

THE PROJECT

3.1   Description of the initial steps to be taken for the implementation of the project.

The first objective of the CPLP Cultural Television Network is to install its structures in the following countries:

-        Sao Tome e Principe
-        Cape Verde
-        Mozambique
-        Guinea-Bissau
-        Angola

-       Timor

Thus, these six countries will structure the basic network of the whole project.

Angola, Mozambique and Cape Verde will form a FIRST GROUP; Sao Tome e Principe, Guinea-Bissau and East Timor will form a SECOND GROUP; and, finally, Brazil and Portugal will form a called REFERENCE GROUP.

However, the project will start at the same time in all countries. A more precise description of the meaning of each group is made below.

The basic physical components of the project are:

1.   The creation of companies, in each of the States of the CPLP, oriented to the objectives above referred.
2.   The State of each country – who will be the responsible and guarantee of the payment of the financing – will also have a participation in the companies.
3.   All strategies will be oriented to lowest costs in long term.
4.   The whole project will be developed by phases.
5.   The timetable of transmission will be defined case by case.
6.   The project will include:

6.1.   a complete plan for an equipment network.

6.2.   a complete plan for a software network.

6.3.   all procurements

6.4.   equipment transportation and installation

6.5.   all configurations

6.6.   creation of the infographic system

6.7.   technical formation of the personal

6.8.   architecture and engineer projects for the building

6.9.   construction of the building organised in phases

1.2   Description of the procedure by which the necessary programmes will be produced for transmission on the channel.

The television programs are known as software in this field. The unit is planning transmission of three hours of educational programs, three hours of cultural programs and two hours of commercial programs per day in the first year. Thus, there shall be a total transmission of eight hours of programs per day in the first year. The total transmission hours shall increase by two hours per day in the second and subsequent years. Thus, in the second year there shall be total transmission of 10 hours per day and in the third year there shall be a total transmission of 12 hours per day. The cost of software production is assumed at US$.2.500 per hour. The amount of US$.2.500 per hour is taken based on cost of production of such programmes in Portugal.

To contract a similar system from other television channel, only for one country, for a transmission period of 6 hours daily, the cost in, market (eg. Portugal*) is of about US$.500.000 per month. This cost represents a total of about US$.6.000.000 per year. In 16 years, excluding the first year of implementation, the costs for contract the six daily hours of transmission from other television channel, in present market price, is of about US$.100.000.000. The six countries would have an estimated cost of about US$.600.000.000 in a 16 years period of time.

These costs are based on the prices presently used in Portugal, as for example the programme Hora Viva – Seguranca em Directo, transmitted every day, excluding weekends, from 7 to 10 AM.

Another example is the channel Canal Noticias Lisboa CNL. Only in its first year of installation, it was expended about US$.10.000.000, and the equipment as well as the building were not property of the channel, but rented.

In fact, however the project will be implemented by phases, after 5 to 10 years of regular work, it is predicted to have reached a similar value in incomes as the examples above.

*value took in the Portuguese market, below the average of the European prices in about 30%.

The first ten months of works will be oriented to:
-        design of the equipment and software structures
-        start the first productions
-        elaboration of the projects of architecture and engineering
-        procurement and acquisition of all equipment
-        construction and supervision of the works
-        installation and configuration of the equipment

So, the first year must be used to the implantation of the project. However, as to accelerate the chronogram, some operations should be started before this timing, temporarily installed in a different building.

The first phase will be oriented to:
-        video productions (external and internal)
-        edition
-        elaboration of cost programmes

3.3   Description of the requirements of Power, Fuel and Water. POWER
It is assumed that the unit shall use one thousand

units of power of everyday. Thus, the unit shall use 365.000 units of power in a year. The costs of power is assumed at five cents for one unit. The consumption of power shall increase by 10% every year.

FUEL

The unit shall use D.G. set for generation of electricity, when there is an interruption in the supply. The hours of disruption in the normal supply can not be predicted accurately. Hence, the hours for which the D.G. sets shall be used also can not be predicted accurately. However, it is assumed that the unit shall utilise one thousand litters of diesel per month. The cost is assumed that 33 cents for one litter. Hence, the total cost for one thousand litters shall be US$.330. The consumption of diesel shall increase by 10% every year.

WATER

The unit does not need water for fixed commercial operation. However, it shall have a strength of 50 people as employees. Water shall be required for drinking and sanitation purposes. It is assumed that 3000 litters of water shall be required everyday. The cost of obtaining this water is assumed at US$.10 everyday. The consumption of water shall increase by 10% every year.

3.4 Description of the manpower requirement of the project and description of the steps taken by the company to train the manpower.

The total staff strength shall be 17 people in the first year.

The system of modern TV Station is designed to achieve sustained operating efficiency and transmission. This, of course, entails a certain degree of sophistication in the production and transmission. However, considering the socio-economic situation in Angola, a reasonable balance has to be struck to obtain optimum performance and at the same time create gainful employment. While working out the manpower requirement for this project to be kept on direct rolls of the company totalling 50 staff and operators, the above consideration have been kept in mind.

Manpower requirement

The direct manpower required for the proposed Unit is about US$.1.800.000 per year. The manpower requirement as indicated in this chapter has been planned keeping in view the following guidelines:

Effective co-ordination among the various departments. Judicious distribution of responsibilities.

Capacity utilisation of the TV Station with optimum manpower.

Details of manpower requirement for the TV Station is given in the financial section.

In line with the prevailing practice, the Security guards, office peons and unskilled labour etc., are normally employed on contractual/casual basis, and their cost has to be included in the Factory overheads. However in this report provision has been made to employ the above staff and their salaried have been included.

The various departments proposed shall be under the direct responsibility of the Station Director. He shall be assisted by Deputy Directors, who will look after the complete TV Station and its various day to day activities and Marketing. They shall be responsible for achieving the envisaged targets and sales forecasts. They shall be assisted by a team of Managers from production, marketing and accounts.

To run this project a labour force which shall usually be composed of unskilled and skilled workers shall be employed. The first are those who do not undergo any king of specific training or education, while the latter have to do so in order to master their jobs.

When evaluating an investment project from an employment point of view, its impact on both unskilled and skilled labour has been taken into account. Not only direct employment, but also indirect employment has been considered. Direct employment refers to the new employment opportunities created within the project; indirect employment concerns job opportunities created in other projects linked with the project which is being formulated.

The implementation of large and sophisticated projects generally contributes to the development of local skills and capabilities in a country. Furthermore, they help to change traditional values, attitudes and the behaviour of the society, to build up an enterprising spirit among the people, to develop a desire for changing and improving the existing conditions of life, to introduce better work discipline and thus to change the very pattern and basis of economic development. The TV industry is already well established in Africa. Location of TV industry activities in Africa, has certain favourable factors and advantages, in setting up this type of industry, as availability of acceptable level of education among the supervisory staff adequate technical and managerial skills developed over a long period, and availability of cheap labour are assured.

The organisation structure will vary from TV Station to TV Station in the industry and as such the pattern proposed herein can be considered only as suggestive and provisional.

A well-knit organisation structure headed by a Station Director and Manager, with the supporting staff will be developed progressively during project implementation. Soon after the plant becomes operative, a good number of project staff will be absorbed in the organisation. Certain additional staff also get added to ensure smooth and efficient management of the operating unit.

The Manager Production will have a degree in communication and will be accountable and answerable to the Station Director for all TV operations including planning, production, material management, TV utilities, quality control, production cost and budgetary control, TV safety, discipline and layout relations. His main function should be to ensure achievement for quality and quantity targets of production at reasonable cost and should constantly strive to improve TV performance. In the discharge of his multifarious duties and responsibilities he will be assisted by supervisors and adequate staff for day to day activities.

His duties are to ensure that targets are maintained through effective utilisation of 3 M's viz: Machinery, Men and Materials. These targets should be translated in terms of targets for individuals under him such as supervisors, skilled workers etc. He has to ensure that the equipment at his disposal gets prompt attention on breakdown is adhered to strictly. He has to further ensure that there is a strict quality control exercised over raw materials also. He will be assisted by Deputy Manager etc. who and will report to the General Manager.

GENERAL MANAGER (Comm.) He will report to the Station Director. His main duties will be prepare sales forecasts and budgets, study and monitor the export market for the company's products and advice on ways and means of increasing sales. He will also ensure that consumer complains are solved in an appropriate manner. He will also ensure that the right materials are available for production at the right time.

He will be responsible for all aspects connected with the export procedures and keep the management updated on all matters relating to the Govt. policies on polymers and Exports and above all the world market.

COMPANY SECRETARY & FINANCIAL CONTROLLER. He will report to the Station Director and will be responsible for the departments of administration, financial planning, budgetary control, cost accounting, tax management payroll accounting, etc. and all personnel matters. He will be assisted in their duties by their respective assistants to assist in day to day activities.

Manpower planning and production

The central issue here may be one of scale. Production is staffed bye personnel, but the process may be labour-intensive or capital-intensive. Either way, planning will include:

1.	Analysis of labour supply and demand factors in relation to skills and training needs.
2.	Procedure for manpower recruitment, together with selection processes.
3.	The formation of industrial relations policies necessary for effective work place bargaining, disciplinary measures and dismissal procedures.
4.	Analysis of the effective use of human resources.
5.	Conditions necessary to maintain adequate levels of motivation. The scale of the problem is likely to be directly proportional to the method of production.

The availability of main persons is not going to be easy since TV industry is currently in its infancy in Africa.

Training needs

The selection and training of the required manpower for the proposed project has to be planned in advance.

The key personnel should be selected and trained suitably. The training would be carried out in the following manner:

Basic training on the concept of TV industry before construction begins with visits to similar TV Stations in other countries. On site training during the construction phase of the project. On job training during the commissioning phase of the project. On job training during operation of the TV immediately after commissioning. The training of the key personnel such as Station Director should be carried out in all the phases. The training of other operating personnel should be suitably carried out during the construction, erection and operation phases in addition to training them by visits to similar plants operating in other countries.

Besides training the key operating staff described above, in TV training should also be given to other employees at skilled operating level to enable them to understand the process equipment in the project and prepare them to operate a maintain their respective sections safely, efficiently and skilfully. The above training should be carried out during construction, commissioning and operating phases of the project.

Training is necessary in order to enable personnel to

acquire the skills and knowledge necessary to perform a task to an acceptable standard. The length of the training period and training methods will, of course, vary from job to job. Training is essentially a learning process, and in order that progress can be successfully monitored certain conditions are necessary.

1.   The training needs of both the individual and the organisation shall be identified and analysed.

2.   Targets and standards shall be set for the trainee, which are within his capabilities.

3.   The pace of the training programme should reflect the trainee's ability to maintain progress in properly absorbing the same.

4.   The trainee shall receive regular feedback of results. Any problem areas shall be highlighted, discussed and resolved.

5.   As the trainee progresses the amount of information provided shall be gradually reduced, thus inducing a feeling of independence and competence.

It is common place to find a wide variety of tasks in an organisation and each will require varying degrees of skill, effort and responsibility. This being so, it is inevitable that rates of pay will also vary and the differentials between the jobs will reflect their relative values. However, other factors such as local market conditions, bargaining strengths and traditions also influence a company's payment structure and a great deal of planning is required if rationalisation is to be achieved. One technique which has been successfully adopted by many companies to establish an equitable wage structure is job evaluation. In a job evaluation exercise a comparison is made of common criteria over a range of jobs, and the resulting analysis may be linked to a points allocation or job ranking system, and hence to a wage scale.

In conducting a job evaluation exercise it is important to cover a reasonable variety of tasks within the whole spectrum. For each, a job description is prepared setting out details of the duties and responsibilities undertaken by the employee together with a statement about his working conditions. Very often this task is undertaken by work study personnel since, they are responsible for determining methods of operation and evaluating the work content of the job. Each job is assessed factor by factor, resulting in a comprehensive comparative analysis.

The individual is the most important resources of any company and only people who are well trained, well motivated and adequately rewarded will provide a positive and synergistic contribution towards the company's objective and its organisational health.

In most cases, the factors, which may be weighted according to relative value, are as follows:

1.   Skill-education, experience and training.
2.   Effort-both physical and mental.
3.   Responsibility-for equipment, materials, initiative etc.
4.   Working conditions-general conditions, risk of accident and injury.

Pay policies affect not only individual employees but the whole organisation, and the rewards and objectives vary at different levels within the enterprise.

Industrial relations

An effective industrial relations policy is important, since is the system through which employees take part in decision-making and in many instances it affects the while atmosphere of employer/employee relationships. An industrial relations policy is essentially a set of rules whose determine procedures for negotiation on such matters as:

1.   Wage and salary scales.
2.   Terms and conditions of work.
3.   Disputes and grievances.
4.   Recruitment and dismissal.

5.   Other issues of mutual interest, e.g. closed shop, redundancies and joint consultation.

In order to promote an atmosphere of co-operation, and to minimise conflict, the needs of management and work people must be recognised by both sides. Trade unions exit to protect the interests of their members and improve their working conditions. Management, while aware of the pressures and constraints imposed by the trade unions, have a duty to maximise the use of resources at their disposal, which may be expressed in relation to profitability, return on investment, level of service, sales volume, market share and cost-effectiveness. The strategies adopted in attempting to solve industrial relations problems will vary from company to company, and indeed from union to union, but there is no doubt that they will be influenced by both internal and external factors.

Internal factors

1.   The attitudes of employees to management, and management to employees.
2.   The leadership style of management.
3.   The bargaining strength of both parties.
4.   The number of negotiating bodies.
5.   The prosperity of the company.

External factors

1.   The extent to which parent boards influence company management, and district officials influence or control local shop stewards.
2.   Whether or not bargaining is conducted at plant, local or national level.
3.   Government policy towards industrial relations.
4.   The economic situations nationally, locally or within the company itself.

3.5   Describe the project implementation schedule.

Implementation of this project is a challenging task and calls for meticulous planning, scheduling and monitoring to realise the project goals within the budgeted cost and time frame. The goal can be achieved by adopting modern project management techniques.

To implement this project adequately, a team of engineers and project personnel having requisite education and experience are being appointed, to whom a detailed Work Breakdown Structure (WBS) in a logical order of activities, shall be supplied shortly, keeping in view cost estimation, scheduling, and to help monitor and control of the project. It is proposed to be formulated in conjunction with the objectives of each activity and goal settings. Project shall be programmed and controlled by network analysis techniques. Before the application of the network analysis techniques, the project personnel shall be acquainted with their capabilities in saving time, resources and costs. The training of project personnel and engineers at levels shall be provided for proper control of project progress and taking of timely corrective actions to re-align these efforts to meet ore stated objectives. It is proposed to gear programming and control system, i.e. project implementation system, which will ensure an integrated approach to project implementation. Project management activities shall be determined in advance and all activities carried out be project personnel as well as those to be contracted shall be identified.

Responsibility for project implementation shall be clearly defined. The forms of project organisation range from project oriented to functional organisation, while most of the cases are combinations of the two, with certain adaptations to prevailing conditions. It is impossible to over emphasise the importance of establishing a team of a task force for implementing the project with a designated leader to co-ordinate and guide its functions.

Project manager: Shall be responsible to the Board of Directors. The project manager shall be responsible for guiding and co-ordinating the efforts of all parties engaged in implementing the project, obtaining necessary government approvals on contracts. He is to control the project organisation with the promoters as well as with other agencies and organisations interested in the project. The manager shall have some staff to assist him, especially in checking expenditures to date and determining the present and future cost overrun or under run so that the project manager can take or propose to the Board pertinent corrective measures.

The network shall cover the pre-construction phase of the project indicating major administrative processes, since experience shows that some of them have frequently involved lengthy delays. In other words, it shall include the aggregate activities to be carried out be the principal parties participating in the implementation process.

The project budget shall then be prepared. Part or periodic payments to contractors which might be made at the end of certain intervals (e.g. weekly or monthly) throughout the time horizon of project implementation, shall be made by summing up activity costs per unit of time, which may be a week & month, and computing the cumulative cost at the end of each time interval. For the activities that are in process and are contacted or sub-contracted, the assumption of a linear time cost activity relationship shall be used for the sake of simplicity. In other words, expenditures are uniformly distributed throughout the duration of the activity.

PROJECT SCHEDULE

After an investment decision is taken, the main machinery and long delivery items must be ordered out at the earliest, forming the first major step in implementation of the project. It is foreseen that an engineering consultant will be appointed for carrying out the detailed engineering including basic engineering and procurement assistance to the client. It is also assumed that reputed and experienced contractors with adequate resources viz., men, materials, tools and tackles etc. will be engaged for execution of the construction and erection work. The purchase packages for auxiliaries shall be kept minimum so as to reduce the co-ordination efforts to the minimum. A great deal of co-ordination is required for constructing/erecting the new units. This task is feasible, provided the major activities of the project are co-ordinated and completed in the duration specified to achieve the respective milestones in time.

STRATEGY FOR TIMELY EXECUTION

It is important to deploy a team of experienced personnel for project execution and select the external agencies with due care for rendering the services and supply of equipment for the project. The project activities must be identified, planned and scheduled, and the progress monitored for timely project implementation. All the inputs to the project including financial resources must be identified and their inflow planned and arranged in time.

The project must be managed professionally with necessary co-ordination among the various agencies and requisite decisions taken promptly.

Establishment of an effective monitoring procedure for progress review and co-ordination.

In short, the following key factors would constitute the broad strategy for timely execution of all activities in a pre-determined manner as per schedule shown in the bar chart as to reach at a basis of regular production.

Early selection of an effective in-house technical team (TASK FORCE) by Government of Angola, headed by a Project Manager for planning and executing the project.

i)   Proper choice of external agencies such as consultants for Project Engineering., Machinery suppliers, Construction Agencies etc. keeping in view their reputation/past performance and working experience in their fields.

ii)   Adequate use of computer-based PERT/CPM techniques for project planning, scheduling and monitoring.

3.6   Description of the initial project Flowchart.

The project implementation phase embraces the period from the decision to start the project to the beginning of the commercial production. It includes a number of stages including negotiations and contracting, project design, construction and start-up.

3.7   Description of the requirement of Land.

For this project the Government of Angola will participate with about 25.000 square meters of land. In selecting the land the following criteria should be followed.

1.   The land should be near to the main city of Luanda.
2.   The land should be properly connected by road.
3.   It should also have proximity to many end user clients.
4.   It should also meet various Government

policy of achieving the social objectives.

2.08 Description of the requirement of Building.

The total requirement of building is as follows.
a. Adm. Building   1000 Square Meters
b. Studios –3 (Three)   2500 Square Meters
c. Other Misc. Building   500 Square Meters

Total Square Meters: approximately 4000 Square Meters.
Cost for construction of about US$.750 per square meter (including air conditioning systems, primary electrical cabin etc.).

2.9   Description of some requirements for Plant & Machinery and their basis of selection.

The plant and machinery have been selected giving due consideration to the sophisticated nature of technology required. Detailed discussions were carried out with the foreign suppliers to ensure that the required capacities are practical with minimum capital and operating costs before the machinery were finally selected.

The machinery shall be selected from the most reputed foreign manufacturers of complete range of
equipment. Keeping in view the following main factors:

a)   Past performance
b)   Existing machinery in Africa & abroad.
c)   Sales and service facilities in Africa

The main equipment is as follows.

STUDIO

1.   Cold lighting system
2.   8 Camera channels composed by :
4 DXC-D30PK Sony Digital Cameras
6 J18 Canon Zoom Lens
8   CA-537P Sony Camera Adapters
8   DXF-50 Sony Studio Viewfinder's
8   CCU-M3P Sony Control Camera units
8   RM-7P Sony remote CCU Controls
8   50M Sony Multicore Camera Sets
8   Canon Focus and Zoom control sets
8   Intercommunication Headphones

8 Special tripods
3 charriot plane and curve
2 Teleprompter with 2 reading systems (monitors for individual speakers)
2 lighting control table with memory-ARRI
Different microphones with diverse directional characteristics Audio Components
Video Components
General Intercommunication system

Video Regie
1.   3 Recorder/Player BETACAM equipment
2.   1 video digital mixer with 12 tracks
3.   1 DVE equipment of digital effects for 3 channels
4.   1 TBC Frame Synchroniser
5.   2 Oscilloscope & Vectorscopy for control of cameras
6.   2 Synchroniser Generators with Changeover
7.   2 Intercommunication system with 4 places
8.   2 Audio & Video Matrix with remote controls
9.   Video distributors

Audio Regie
1.   1 audio mixer with 24 tracks
2.   2 audio monitors
3.   2 cd player

4.   2 cassette deck
5.   2 DAT
6.   2 Mini-Disc.
7.   1 Level Detector for Audio Stereo
8.   Stereo Audio Distributors
9.   Audio compressors
10.Audio effects 11.Microphones

Video Post-Production
1.   6 Sony Es-7 on DVCAM/Avid hybrid non linear edition stations

2.   2 Conventional BETCAM Edition Suite 2:1
3.   Multi-Track audio (8) on hard disc, sound scape type

3.   ENG SETS

1.   Sony DSR-300 PK DVCAM Camcorder Digital Compact
2.   Report projectors
3.   Various support materials(Batteries etc)

3.10   Description of the Water Requirements for the project.

The requirements of water, separately for various matters are given in the following table.

Circulating   : Nil
Make-up   : Nil

Process   : Nil
Drinking   : 3,000 LPD

3.11 Description of the Steam requirement for the project.

A. Steam requirements and steam balance   : N.A.

B. Capacity and type of boiler with detailed specifications   : N.A.

C. Steam and energy diagram : N.A.

D.   Total energy generated / purchased (converted into M. K. : N.A. Cal) theoretical requirement of energy (in M. K. Cal)

at the various consumption stations and expected actual requirement at these stations.

E. If alternate processed are available, comparative energy : N.A.
consumption figures for the various processes. If the
project is energy intensive, possibility of choosing alternate process in order to make the project less energy
intensive.

F. Steps proposed to be taken by the company to improve : N.A.

energy losses efficiency and reduce energy losses (such

as power factor improvement, power load management,

optimising, illumination waste heat utilisation, etc.)

G. Scope for usage of solar / other renewable sources of energy: N.A.

H. Any other measures contemplated in the direction of : N.A.
energy conservation and management.

3.123 Some information on Compressed air, fuel, etc.

Compressed Air

(a) Requirement   : N.A.
(b) Sources   : N.A.
(c) Arrangements proposed   : N.A.
(d) Cost at site with detailed calculations : N.A.

3.13   Details of the nature of atmospheric, soil and water pollution likely to be created by the project and the measures proposed for control of pollution. Indicate whether necessary permissions for the disposal of effluent have been obtained.

There shall not be any atmospheric pollution likely to be created by the project as there is no machine which has combustion resulting into air pollution or any chemical process which may release any gases which may result into an air pollution.

However, the use of DG set shall cause a small amount of air pollution. Considering the size of DG set, the pollution is within the permissible limits.

ANGOLA TELEVISION NETWORK

THE INDUSTRY

CHAPTER - 4

THE INDUSTRY

4.01 Description of the Television Industry in general.

The television industry can be described in the following broad categories. 1. Introduction
2.   Cable Networks

3.   DTH

4.   TV-media

5.   Earnings drivers

6.   Outlook

The detailed discussion in each of the categories is given below.

Introduction

The growing popularity of TV as a communication medium has resulted in the TV media sector undergoing a rapid transformation. From the black and white days of state controlled TV Station, to the highly colourful tunes of Channel V and MTV, the medium has certainly undergone a phenomenal change. Given its popularity, percentage ad spend has also increased proportionately on this medium.

Media pie (%)

	1995	1997
TV	62.5	68.8
Radio	20.9	15
Press	16.6	16.2

Source NRS

Entry of new channels post 1991

All over the world the telecasting has witnessed entry of new channels to cater to the various needs of world audiences. Channels have been launched in English as well as other regional languages. In many countries of the world till 1991, the state owned TV Station ruled the roost, as other players were not allowed to uplink and broadcast. However channels such as CNN, Star TV and BBC, which were offshore companies, could circumvent these regulations and telecast their programs into any country of the world. Cable operators then relayed the same and made it available to the common man through the cable television network.

Like many other countries, the State machinery controlled television. It was used as a propaganda tool for the party in power, with the opposition always at the receiving end. The customer had very little choice. The first steps towards more user choice began during the 1980s, which had to be telecast to a wider audience. TV Station used satellite channels for the telecast and the T.V. network was launched as an international channel.

The sports telecast by Channel 9 in 1985 and the Gulf War in the late eighties all played small but important cameos in educating the international viewer. With liberalisation in 1992 and crumbling tariff barriers televisions (read as colour TVs) became more easily available. The media revolution had started.

Major satellite channels avidly watched by viewers are Star TV, Sony TV, Home TV, BBC & CNN. There are other regional language channels which are major players in their respective territories.

Most of the channels that could not attain popularity rapidly suffered, since their advertisement earnings were not sustainable. The first round of the media wars is over. Management changes, i.e. original promoters selling out to new management with deeper pockets, has become the order of the day. Alliances like the famous ESPN Star Sports arrangement also made headlines. Given the global trends of mergers and acquisitions, further consolidation is likely. Alliances and mergers make sense when the partners complement each other, like BBC and Discovery launched Animal Planet, CNBC and ABNI came together to launch a business channel called CNBC Asia.

Cable Networks

Antennas set up by either the end user or the cable operator receives the signals transmitted by the satellite. Local cable operators lay their own cables, set up control rooms, which can telecast 40 or more channels over a limited area. They charge the household a one- time connection charge of about US$.10 per point and a recurring monthly charge ranging between US$.1 to US$.5.

Initially, this was done in a very unorganised manner. The business required local knowledge and contacts, so every locality had its own cable operator. Collection was critical for the cable operator. For the end user, quality of telecast and a complete lack of standards became an issue. This lead to a shakeout and the formation of cable companies with money power which in turn tied-up with the local and small cable operators. Cable companies charge about US$1 per month to the local cable operators and support them with training and other infrastructure inputs. The business is immensely capital intensive and takes a long time to break even.

In many countries the operations of cable operators are regulated under the Cable TV Act which ensures that pornographic materials and other materials which are against culture and values or are detrimental to national interests do not get telecast. Recently this act has been amended to include foreign channels also.

Direct To Home

DTH is a new technology that circumvents the cable operators by directly delivering a bundle of channels to the end user. DTH involves transmission of encoded audio/ video signals (Ku band) via satellite. The end user needs an antenna to receive the signals and a decoder (set top box) to unscramble the encrypted signals. DTH services elsewhere in the world are Echostar and DirecTV (USA) and BskyB (Europe). Rupert Murdoch of Star TV fame owns BskyB.

The size of the antenna in DTH will be 1.5-2 ft in diameter, making it easy to install and transport. In conventional cable, since signals are in C band, an 8ft- diameter antenna is needed. The basic difference in the business model is the hardware costs in DTH. In a cable system, the user pays a one time connect fee and monthly rentals, while in DTH he has to invest in hardware.

The antenna will cost about US$200-300 and decoder will cost about US$200. The African viewer might be reluctant to incur such heavy installation costs. Quality of telecast in DTH is superior to Cable TV and viewer can receive up to 200 channels.

DTH will result in restructuring of the cable television industry. It will become imperative to have cash reserves to withstand the technology threat. Up gradation to fibre optic backbone will become necessary. A fibre optic network will cost about US$0.5mn per km as compared to US$0.1mn per km for coaxial cable. The stage is now ripe for consolidation.

TV Media features

In the Broadcasting business, it is only the industry leader who makes sizeable profits. The business is a game of asymmetrical payoffs. For instance, the top 5 channels account for 90% of ad spend.

Urbanisation and TV penetration is related. This may be due to the popularity of cable television that has resulted in increased colour TV sales. Rural penetration is low, although growing at a fast pace, because of dearth of specific program content to cater to that segment.

Liberalisation has resulted in the world viewer becoming more aware and conscious. This has resulted in the customer having more choice with the entry of a number of companies in different segments. Competition has resulted in companies increasing their marketing spend significantly.

Popularity of TV media is becoming higher. Increasing TV penetration leads to a reallocation of advertisement budgets with higher allocation for television at the cost of other medium.

TV channel operators use different business models to generate revenues. The critical component of any channel is the quality and type of programs they telecast. This determines their popularity, which in turn determines amount of advertisement revenues they can generate. They can do any one of the following:

Buy programming rights of program software from outside and collect advertisement revenue on their own. This model is followed by several TV companies, wherein they have a separate company in their fold, which develops all the content. The advantage is that re runs of serials/ programs become very profitable.

Selling time space to the producers for a fixed charge. Producers in turn are free to book advertisements at their own rates (there is an understanding on the time allocated for advertisement) and collect revenue. This is the basic model for many TV companies in which they sell prime time slots. The rights continue to be vested with the producer.

Earnings drives

The key factors that drive sector revenues are

Television penetration: Since the medium is television, increased television penetration will imply higher viewer ship. This will translate into higher advertisement spend allocation. This will also imply higher software production and demand for new programs.

Competition from other satellite channels would have an adverse impact on advertisement revenues, as advertisers have more choice in allocating ad budgets.

Government policies can have a big impact on the fortunes of the entire industry. When the DTH bill is passed in any country then, it will trigger a restructuring of the cable business.

Launching new channels targeted at specific segments, like regional channels within any country other areas having large pockets of ethnic population would lead to revenue growth. This will entail significant initial outlays.

Depreciation of the local currency would increase revenues as most of the program/ software companies export the programs overseas and payments are dollar denominated.

Advertisement revenue

As mentioned earlier, this is the primary source of income for TV channel operators. This revenue is directly co-related with the reach and viewer ship of a channel. Any channel's popularity depends on good quality programs, which is the software content. The business requires enormous initial investment in programs and revenues follow only with a time lag after the channel receives a minimum viewer acceptance.

Outlook

The sector has latent potential for growth on back of the exponential growth of cable TVs during the last 5 years. Television penetration in Africa is extremely low as compared to other developing countries like Malaysia, Pakistan, etc. in Asia. The number of channels has increased, implying higher demand for software programs.

Advertisement revenues, which are the barometer of channel popularity, will get dispersed over several competing channels. A shakeout is likely in both the channel and cable TV sectors. The biggest beneficiaries will be the content providers or the software houses. They will control the intellectual rights to the key element driving any channel's popularity.

Direct-to-Home, Digital Terrestrial Transmission and Conditional Access Cable Delivery have emerged as new delivery mechanisms. Breakthrough in technology would help open up avenues for these channels.

ANGOLA TELEVISION NETWORK

MARKETING

CHAPTER - 7

MARKETING

7.1   Description of the commercial viability of the project with regards to revenue generations.

Estimations shows that, after about 5 years of regular working, the CPLP Cultural Television Channel will be able to start exporting a great quantity of services and videos to Africa, Europe and South America. (See financial projections)

The economic feasibility of the project will occur with the commercialisation of several products, like:

1.   The sale and production of cultural programmes to various institutions like:
-       BBC
-       TV Culture Brazil
-       RTP
-       Discovery
Etc.
All over the world commercial networks for television cultural programmes have been created.
Several television channels in different countries
have regularly acquired programmes by producers spread out all over the world. Not only, several
international organisations, like the World Bank
and the United Nations, including FAO and Unesco,
among many others, need programmes for public
education, like programmes oriented to alert and
to educate people concerning endemic and epidemic diseases. AIDS, malaria, tuberculosis are a
very few examples.

These world institutions have made a great effort
to develop television programmes with humanitarian objectives. But, in Africa – undoubtedly the continent most needed of such programmes – there is
no television station, in present times, with capability to make face to such a need.

Therefore, the television teams that had been responsible for such programmes are, practically in

its totality, placed in countries of the called First World, strange to the local population's concrete reality.

Programmes focusing new agricultural techniques or even oriented to agricultural, commercial and industrial education are essential elements of such a repressed demand.

The price for each institutional campaign, with an average of 10 films produced per campaign, is of about USD.15.000$ and the capacity of the CPLP Cultural Channel will be as follows:

First Group* - from 2 to 4 campaigns per month.
Second Group – from 1 campaign in two months to 1 campaign per month

This represents:
First Group   a potential annual income of about US$.540.000
Second Group a potential annual income of about US$.90.000

*First Group: Angola, Cape Verde, and Mozambique. Second Group: Guinea-Bissau, Sao Tome e Principe, East Timor (Lorosae). Reference Group: Brazil, Portugal.

2. The sale and production of no cultural pro-

grammes to other television channels, like:
-        series (novels)
-        talk-shows Etc.

Many countries of the region need to produce television programmes, but do not have capability to do it. Thus, they seem themselves obliged to search expensive productions in Europe. With only one programme sold-three daily hours-the channels of the First Group* would receive in incomes the equivalent of about two times of the whole investment. The Second Group countries will not have conditions at the beginning to product cultural programmes to other television channels.

*First Group: Angola, Cape Verde, and Mozambique. Second Grope: Guinea-Bissau, Sao Tome e Principe, East Timor (Lorosae). Reference Group: Brazil, Portugal.

The price for this type of television programme is:

Auditory programs   US$.12.500 per programme
Talk-shows   US$.4.000 per programme
Interviews   US$.4.000 per programme

The capacity of production, of these programmes by the countries of the First Group will be:

In the FIRST PHASE:

Talk-shows   2 per month
Interviews   4 per month
In the FINAL PHASE:
Auditory programmes   6 per month
Talk-shows   20 per month
Interviews   20 per month

Thus, the potential income will be:

FIRST PHASE
Talk-shows   US$.96.000 per year
Interviews   US$.192.000 per year

FINAL PHASE
Auditory programmes   US$.900.000 per year
Talk-shows   US$.960.000 per year
Interviews   US$.960.000 per year

The price of the novels is much higher. For each complete novel, with about 120 chapters, the price is about US$.2.500.000 and the capacity of the First Group of the CPLP Cultural Channel (in the final phase) will be of one novel per year.

3. There is the possibility of commercial use of the transmission time beyond the six hours reserved to culture and education.

The conventional day in television is of 18 hours, of which only 6 hours would be no commercial. We would have, therefore, 12 hours of no cultural transmission, which should be freely commercialised.

Even the period of six hours- divided into two sections of three hours, the first one dedicated to culture and second section of three hours to education-has a great potential for sponsoring.

Each commercial hour of transmission can include up to 12 minutes of advertising. The price for each 30 seconds of advertising is:

Noble time   from 6PM to 11PM   a b o u t US$.400 each 30"
Normal time   rest   about US$.200 each 30"

Concentrating the commercial programming in the noble time. With 4 hours of transmission in this period, we would have:

Educational television: from 7AM to 9AM
From 3PM to 4PM
Cultural television: from 4PM to 7PM
Commercial television: from 9AM to 3PM
From 7PM to 1AM

Thus, the commercial period would comprehend 6 hours in normal time, 4 hours in noble time and 2 hours in normal time again.

Begin 12 minutes per hour for advertising, the commercial period could have:

Noble time per day	48 minutes	24	films	US$.9.600
Normal time per day	96 minutes	48	films	US$.9.600
Total of the potential income...........				US$.19.200 per day

> or US$.576.000 per month
> or US$.6.912.000 per year

It is believed that in the first phase of the project (after one year), the CPLP Cultural Television Channels of the First Group will be able to start with an income from commercial advertisement of about US$.1.000.000 per year-value which is predicted to increase in the follow months.

4. The rent of the studios to television teams of other countries can be another source of incomes. Many producers who cover events in Africa need to move many times to their countries of origin during the video works, because there is not technical support in Africa. The same phenomenon happens with the cinematography and the journalism productions. It is not difficult to imagine, for example, the serious problems journalism teams have had, for example, with essential components like batteries, lighting, electronic components etc., which only can be easily find in Europe.

The period of renting of a television studio is of 12 hours. Each period has price of about US.4.000. After the first year, the studio should be rented for 12 hours per each period of 3 days. So, the rent per month, in this period (First Phase), will be able to generate incomes of about US$.120.000 per month in the First Group.

5. Support services to other television networks in all areas, including novels, mini series, docdramas series etc.

Many countries, principally in Africa, do not have technical conditions to develop this kind of programmes, but they have a strong internal repressed demand in this sense.

Each team, abroad, has a price of US$.1.000 per day (two people), after the costs of dislocation, hotel and meals. In the first months (first phase) it will be organised one team for works in other countries. In the final phase it is predicted to have up to 5 teams with such function. The capacity of each team is of about 20 days per month. Therefore, the support to other television networks-when took in its full potentiality-will be able to represent up to 20 days per month in the first phase. This represents a potential income of about US$.240.000 per year, and 100 days per month of works in the final phase, signifying about US$.1.200.000 per year of incomes, always referring to the First Group countries.

6. Colloquies, seminars, videoconferences and meetings of different natures.

The building of the CPLP Cultural Television Network will have a medium size auditorium, (First Group), with all conditions to receive seminars, colloquies and meetings of the most diverse nature. Such seminars, videoconferences etc. are important not only for the increase of the incomes, but also to attract specialists of the most different areas, being an important element for the development of all region as well as for the diffusion of the local, regional and continental cultural values.

The meetings, seminars and colloquies attract, in average, about 300 people per event. The price-excepting meals, hotels and transportation-per each participant is of about US$.50 per day. It will be a capacity for up to four events of this type per month, what could represent an income of about US$.60.000 per month or US$.720.000 per year (final phase-after 5 to 7 years).

The project should also turn possible.
-        classes of professional formation in diverse disciplines
-        support for a multimedia high technology centre (First Group)
-        support for a multimedia high cultural centre (Final Group)

ANGOLA TELEVISION NETWORK

PROFITABILITY & CASH FLOW

CHAPTER - 8

PROFITABILITY AND CASH FLOW

8.1   Estimation of cost of production and working results for the first five years of operation.

The estimated cost of production of working results for the first five years of operation are given in the chapter "Financial Projection".

8.2   Cash flow statement for the company as a whole, for five operating years of the project based on the estimates of working results.

A detailed cash flow statement for the company as a whole for five operating years is given in the chapter of "Financial Projection".

8.3   Projected balance sheet for five operating years for the company as a whole.

The balance sheet for five operation years for the company as a whole is given in the chapter of "Financial Projection".

ANGOLA TELEVISION NETWORK

ASSUMPTIONS

CHAPTER - 9

ASSUMPTIONS

1.   The Angola Television Network shall generate the income from the following sources.

a)   Sale of cultural programmes.
b)   Sale of commercial programmes.
c)   Sale of advertisement time during transmission of TV programmes

d)   Hiring of studios.
e)   Supply of technical services
f)   Hiring of conference hall

The detailed assumptions for each of the above mentioned activities are as follows:

a)   Sale of cultural programs:

I.   It is assumed that four cultural programs shall be produced per month in the first year, which can be sold to other countries. This

figure will increase to 5 programs per month in the second year, six programs per month in the third year and so on.

II.   It is assumed that the selling price shall be US$.150 per program.

b)   Sale of commercial programs:

I.   The commercial programs consist of talk shows, interviews and auditory programs.

II.   It is assumed that two talk shows shall be produced per month in the first year, three talk shows shall be produced in the third year, four talk shows per month shall be produced in the third year and so on.

III.   It is assumed that the selling price of one talk show program shall be US$.4000.

IV.   It is assumed that four interviews shall be produced per month in the first year, five interviews shall be produced in the third year, six interviews per month shall be produced in the third year and so on.

V.   It is assumed that the selling price of one interviews program shall be US$.4000.

VI.   It is assumed that no auditory programs shall be produced in the first and the second year. Only when the people have two years of experience then in the third year one auditory program shall be produced per month. In the fourth year two programs shall be produced per month and so on.

VII.   It is assumed that the selling price of one auditory program shall be UD$.12.500.

c) Sale of advertisement time during the transmission of T.V. programs:

I.   There shall be a total transmission of 8 hours per day in the first year. It will go on increasing by two hours in the second and subsequent years.

II.   Out of which advertisement shall be available for programs with transmission period of four hours in the first year. The programs in

which advertisements could be available shall increase by two hours per year in the second and subsequent years.

III.   In one hour T.V. transmission, 12 minutes of advertisement shall be allowed.

IV.   Out of total T.V. time 33 % time shall be considered as prime time and remaining 67% shall be considered as non prime time, in the first, second and third years. From the fourth year onwards the prime-time advertisements shall remain constant.

V.   The advertisement rates shall be US$ 800 for one minute of prime time advertisement and US$ 350 for one minute of non prime time advertisement.

d)   Hiring of studios:

I.   It is assumed that the studios shall be taken on hire for a shift of 12 hours, two times in a month in the first year. It will increase to

three times in a month in the second year.
II.   The rent per day shall be US$.4.000.

e)   Supply of technical services:-

I.   It is assumed that one team, consisting of two technically qualified people shall be available in the first year. In the second year two such teams shall be available. In the third year three such teams shall be available and so on.
II.   It is assumed that in the first year the team shall be hired for 10 days in a month. In the second year the teams shall be hired for 12

days in a month. In the third year the teams shall be hired for 14 days in a month and so on.

III.   The rate of hire shall be US$.1.000 per team per day.

f)   Hiring of conference hall :-

I.   It is assumed that the conference hall shall be taken on hire for two times in a month in the first year, three times in a month in the

second year and four times in a month in the third year and so on.

II.   The hire charges shall be US$.500 per conference.

2.   The unit is planning transmission of three hours of educational programs, three hours of cultural programs and two hours of commercial programs per day in the first year. Thus, there shall be a total transmission of eight hours of programs per day in the first year. The total transmission hours shall increase by two hours per day in the second and subsequent years. Thus, in the second year there shall be total transmission of 10 hours per day and in the third year there shall be a total transmission of 12 hours per day. The cost of software production is assumed at US$.2.500 per hour.

3.   The cost of miscellaneous consumable

items is assumed at US$.36.000 in the first year. This will increase by 10% in the second and subsequent years.

4.   The cost of stores and spares consumed is assumed at US$.42.000 in the first year. It will increase by 10% in the second and subsequent years.

5.   The cost of repairs and maintenance is assumed at 0.1% of the total cost of plant and machinery in the first year. This cost will increase up to 0,3% in the fifth year.

6.   It is assumed that the unit shall use 1.000 units of power every day or 365.000 units of power in the first year. The consumption of power shall increase by 10% in the second and subsequent years. The cost of power is assumed at 5 cents for one unit.

7.   It is assumed that the unit shall consume 1.000 litters of diesel per month. The cost is assumed at 33 cents for one litter. It will increase by 10% in second and subsequent years.

8.   The unit shall have a technical staff of 35 people and administrative staff of 15 people.

9.   Depreciation is calculated on reducing bal-

ance method. The rate of depreciation is taken at 10% for building, 25% for plant and machinery, 10% for furniture and fixtures and 20% on vehicles.

10.   Advertisement expenditure is assumed at 0.2% of the sales.

11.   It is assumed that the stock of raw material and stores shall be 30 days of consumption, receivables shall be 45 days sales and creditors shall be 15 days purchases.

12.   The term loan shall be for a total period of 25 years. It shall have a moratorium of five years. It shall be repayable in 20 equal yearly instalments.

PROJECT REPORT

OF

CAPE VERDE CULTURAL TELEVISION NET-WORK

AT

CAPE VERDE

AFRICA

PROMOTERS

REPUBLIC OF CAPE VERDE

AFRICA

CAPE VERDE TELEVISION NETWORK

INTRODUCTION

CHAPTER - 1

INTRODUCTION

1.1   Description of the organisation through which this project is being taken up.

The CPLP is an international organisation, created in 17 of July of 1996, with Headquarters in Lisbon and consisting of the following State Members: Angola, Brazil, Cape-Verde, Guinea-Bissau, Mozambique, Portugal, Sao Tome e Principe and East Timor (Timor Lorosae).

The CPLP has as its objective the politic-diplomatic relationships between its Members, mainly referring to the United Nations and the World Bank, the relationship EU/MERCOSUL and the accomplishment of the Europe-Africa Summit.

The CPLP is also dedicated to the co-operation, particularly in the economic, social, cultural, legal and technical-scientific matters; as well as to the projects of promotion and broadcasting of the Portuguese Language, nominated to the improvement of the International Institute of the Portuguese Language and the creation of a Bibliographical Fund.

All its decisions are taken by consensus.

2.2   Description of the main reasons for setting up an African Cultural Television Network.

Africa is a continent submerged into dramatic problems. All these humanitarian problems have in the education and culture their most direct and objective root.

The African countries speaking Portuguese are among the African countries with the lowest level of development.

The whole situation is so dramatic that the implementation of the CPLP Cultural Television Network is more than urgent.

It is hoped that with this Television Network the people shall have better understanding of various diseases and other problems of life. e.g. AIDS is prevalent disease in Africa. But as the people are illiterate the information about the disease cannot be spread by any other method. In such cases the Television Network becomes a very effective audio-visual means of communication.

The project also intends to be a base of a massive production of health campaigns, specially to reinforce the combat against the AIDS, the infect-contagious diseases- endemic or epidemic-through a process of global education of the local and regional populations.

The CPLP Cultural Television Network is a project for peace and its objective is not to compete with local television channels already installed, but to interchange with them, promoting a future global web dedicated to knowledge and development.

1.03 Description of the main objects of the project.

The major objective of the project is to assist the CPLP countries to develop the first African television network channel oriented to cultural and educational themes with the follow orientation:

-        diffusion of the Portuguese language
-        education in all areas
-        reinforcement of the local and regional cultures
-        formation of technical personal on communication and tele-education
-        formation and information for a better participation of the CPLP nations in the process of global development

-        to support the implantation of the Institute of Portuguese Language

The development of such a project will create an irradiation pole from the countries with the project implementation to the whole African continent, constituting an African network of research and culture, discovering and valorising different cultural aspects and, consequently, reinforcing identity factors.

With the stronger cultural identity, it is expected to have lower levels of violence in social behaviour as well as a relevant increase of the educational performance in all its sectors.

Systematic educational objectives are complementary to the general orientation of this project, which is cultural in its fundament.
A very large positive impact is also to be expected with respect to several sectors of the economy, greatly improving the overall quality of services.

1.4   Description of the mission status of the project.

The mission of the project is:

-        to rescue lost values and cultural information
-        to reinforce historical elements

- to rescue old artworks
- to give a new value to new artworks
- to communicate practical and objective information on health
- to communicate practical and objective information on agriculture
- to communicate practical and objective information on economy

1.05 Description of the philosophy of the project.

The CPLP Cultural Television Network will have in its philosophy, as a support to its mission, the follow fundaments:

- to work with local people
- be focused on quality
- intercommunication

Three words can define the CPLP Cultural Television Network philosophy:
People, quality and intercommunication

Following these three key words of its philosophy, a single sentence also defines its mission: "A Continent of Culture".

Two main sectors characterise its global structure:
1. Cultural Sector
1.1.   To develop programmes on:
- Music

- Architecture (also vernacular)
- Archaeology
- Literature and poetry
- History
- Tourism
- Dance
- Plastic arts
- Astronomy
- Science in general
- gastronomy
- health
- anthropology
- local cultures
- others

1.2 The programmes can be of all types, the most important is to be attentive to communication. All programmes must reach the highest possible number of spectators.

2. Educational sector.
2.1 Systematic education in short practical courses.
2.2 Educational series with practical and objective information on:
- health care
- family
- agriculture
- managing
- services
- civil construction

As the CPLP Cultural Television Network must not be supported by the respective governments in the future, it will have a daily time of free commercial programmes up to 12 hours. This is a good solution to pay its costs and support its development.

1.06 Description the magnitude of the project.

The project proposes a Television transmission time of eight hours per day in the first year. It will increase by two hours per day in the second and subsequent years.

The project also envisages training of local people in all the related areas of Television. It is proposed to develop a local pool of talent which can be utilised not only in this particular country but also in other countries of Africa.

The total cost of the project of the Cape Verde Cultural Television Network is US$.18.400.000. Out of this it is proposed to obtain a soft loan of US$.15.400.000. The remaining amount of US$.3.000.000 shall be contributed by the Government of Cape Verde in terrain for the installation of the respective building, infrastructures and logistic support.

For the purpose of the Cape Verde Cultural Television Network a company shall be formed with the participation of the Government. The Contribution of the Government in the form of share capital shall be US$.3.000.000.

The term loan shall be on soft terms. The total period of the loan shall be 23 years. Out of which five years shall be a period of moratorium. The entire loan shall be repaid in the remaining 17 years, in 17 equal yearly instalments.

The Government of Cape Verde will give the sovereign guarantee for the referred loan.

CAPE VERDE TELEVISION NETWORK

THE COUNTRY

CHAPTER - 2

THE COUNTRY

2.1 - Description of the level of development of Cape Verde.

Cape Verde's low per capita GDP reflects a poor natural resource base, serious water shortages exacerbated by cycles of long-term drought, and a high birth rate.

The economy is service-oriented, with commerce, transport, and public services accounting for almost 70% of GDP.

Although nearly 70% of the population lives in rural areas, the share of agriculture in GDP in 1995 was only 8%, of which fishing accounts for 1.5%. About 90% of food must be imported. The fishing potential, mostly lobster and tuna, is not fully exploited. Cape Verde annually runs a high trade deficit, financed by foreign aid and remittances from emigrants; remittances constitute a supplement to GDP of more than 20%. Economic reforms, launched by the new democratic government in 1991, are aimed at developing the private sector and attracting foreign investment to diversify the economy. Prospects for 1998 depend heavily on the maintenance of aid flows, remittances, and the momentum of the government's development program.

OFFICIAL NAME: Republic of Cape Verde
CAPITAL: Praia
SYSTEM OF GOVERNMENT: Multiparty Republic
AREA: 4,033 Sq Km (1,557 Sq Mi)
ESTIMATED 2000 POPULATION: 437,500

LOCATION & GEOGRAPHY : Cape Verde is an island group or archipelago of ten islands and five islets located in the Atlantic Ocean off the west coast of Africa. The archipelago is divided into two groups (1.) the Barlayento or windward islands in the north and (2.) the Sotavento or leeward islands in the south. The islands are of volcanic origin and except for the Boa Vista, Maio and Sal are mountainous with rugged cliffs and steep ravines.

The coastal plains are semi desert with fine sandy beaches while the mountains are covered by thin forests. Major Cities (pop. est.); Praia 61,700, Mindelo 47,100, Sao Filipe 5,600 (1990). Land Use; forested 1%, pastures 6%, agricultural-cultivated

11%, other 82% (1993).

CLIMATE: Cape Verde has a tropical climate with two seasons. A cool dry season from December to June and a warm season between July and November. Rainfall is low and unreliable with most of it occurring during August and September. The islands suffer from severe shortages of water and rainfall which cause catastrophic and prolonged droughts periodically. Tropical heat and high humidity prevail throughout the year and the conditions are uncomfortable except when fanned by the northeast sea breezes. Average temperature ranges in Praia are from 19 to 25 degrees Celsius (66 to 77 degrees Fahrenheit) in February or March to 24 to

29 degrees Celsius (75 to 84 degrees Fahrenheit) in October.

PEOPLE: Around 71% of the population are Creoles of mixed Black African and Portuguese descent while the remainder are almost all Black Africans with a small number of Whites.

DEMOGRAPHIC/VITAL STATISTICS: Density; 85 persons per sq km (219 persons per sq mi) (1991). Urban-Rural; 29.7% urban, 70.3% rural (1990). Sex Distribution; 48.0% male, 52.0% female (1990). Life Expectancy at Birth; 60.0 years male, 64.0 years female (1992). Age Breakdown; 45% under 15, 31% 15 to 29, 10% 30 to 44, 7% 45 to 59, 7% 60 and over (1990). Birth Rate; 48.0 per 1,000 (1992). Death Rate; 10.0 per 1,000 (1992). Increase Rate; 38.0 per 1,000 (1992). Infant Mortality Rate; 61.0 per 1,000 live births (1992).

RELIGIONS: Mostly Christians with the Creoles and the majority of the Black Africans of the Roman Catholic faith which accounts for around 98% of the population.

LANGUAGES: The official language is Portuguese, although the national language is a Portuguese Creole known as Crioulo.

EDUCATION: Aged 25 or over and having attained: no formal schooling or incomplete primary 84.2%, primary 12.4%, secondary 1.7%, higher 0.5%, not specified 1.2% (1980). Literacy; literate population aged 15 or over 73,500 or 47.4% (1985).

CURRENCY: The official currency is the Escudo (CVEsc) divided into 100 Centavos.

ECONOMY: Gross National Product; USD $347,000,000(1993).PublicDebt;USD$148,000,000 (1993). Imports; CVEsc 12,387,000,000 (1993). Exports; CVEsc 312,200,000 (1993). Tourism Receipts; N/A. Balance of Trade; CVEsc -12,075,000,000 (1993) . Economically Active Population; 120,565 or 35.3% of total population (1990). Unemployed; 25.8% (1990).

MAIN TRADING PARTNERS: Its main trading partners are Portugal, Spain, France and the Netherlands.

MAIN PRIMARY PRODUCTS: Bananas, Basalt, Beans, Coffee, Kaolin, Livestock, Maize, Pozzuolana, Salt, Sugar, Sweet Potatoes, Yams.

MAJOR INDUSTRIES: Agriculture, Canning, Cement, Fishing, Food

Processing.

MAIN EXPORTS: Bananas, Coffee, Fish, Salt.

TRANSPORT: Railroads; nil. Roads; length 5,615 km (3,489 mi) (1987). Vehicles; cars 13,027 (1988), trucks and buses 4,356 (1988). Merchant Marine; vessels 40 (1990), dead weight tonnage 29,730 (1990). Air Transport; passenger-km 122,959,000 (76,403,000 passenger-mi) (1987), cargo ton-km 2,345,000 (1,606,000 short ton-mi) (1987).

COMMUNICATIONS: Daily Newspapers; nil. Radio; receivers 57,000 (1994). Television; receivers 5,000 (1987). Telephones; units 15,000 (1994).

MILITARY: 1,100 (1994) total active duty personnel with 90.9% army, 0.0% navy and 9.1% air force while military expenditure accounts for 1.0% (1992) of the Gross National Product (GNP).

CAPE VERDE TELEVISION NETWORK

THE PROJECT

CHAPTER - 3

THE PROJECT

3.1   Description of the initial steps to be taken for the implementation of the project.

The first objective of the CPLP Cultural Television Network is to install its structures in the following countries:

- Sao Tome e Principe
- Angola
- Mozambique
- Guinea-Bissau
- Cape Verde
- Timor

Thus, these six countries will structure the basic network of the whole project.

Cape Verde, Mozambique and Angola will form a FIRST GROUP; Sao Tome e Principe, Guinea-Bissau and East Timor will form a SECOND GROUP; and, finally, Brazil and Portugal will form a called REFERENCE GROUP.

However, the project will start at the same time in all countries. A more precise description of the meaning of each group is made below.

The basic physical components of the project are:

1.   The creation of companies, in each of the States of the CPLP, oriented to the objectives above referred.

2.   The State of each country – who will be the responsible and guarantee of the payment of the financing – will also have a participation in the companies.

3.   All strategies will be oriented to lowest costs in long term.

4.   The whole project will be developed by phases.

5.   The timetable of transmission will be defined case by case.

6.   The project will include:

6.1.   a complete plan for an equipment network.

6.2.   a complete plan for a software network.

6.3.   all procurements

6.4.   equipment transportation and installation

6.5.   all configurations

6.6.   creation of the infographic system

6.7.   technical formation of the personal

6.8.   architecture and engineer projects for the building

6.9. construction of the building organised in phases

1.2 Description of the procedure by which the necessary programmes will be produced for transmission on the channel.

The television programs are known as software in this field. The unit is planning transmission of three hours of educational programs, three hours of cultural programs and two hours of commercial programs per day in the first year. Thus, there shall be a total transmission of eight hours of programs per day in the first year. The total transmission hours shall increase by two hours per day in the second and subsequent years. Thus, in the second year there shall be total transmission of 10 hours per day and in the third year there shall be a total transmission of 12 hours per day. The cost of software production is assumed at US$.2.500 per hour. The amount of US$.2.500 per hour is taken based on cost of production of such programmes in Portugal.

To contract a similar system from other television channel, only for one country, for a transmission period of 6 hours daily, the cost in, market (eg. Portugal*) is of about US$.500.000 per month. This cost represents a total of about US$.6.000.000 per year. In 16 years, excluding the first year of implementation, the costs for contract the six daily hours of transmission from other television channel, in present market price, is of about US$.100.000.000. The six countries would have an estimated cost of about US$.600.000.000 in a 16 years period of time.

These costs are based on the prices presently used in Portugal, as for example the programme Hora Viva – Seguranca em Directo, transmitted every day, excluding weekends, from 7 to 10 AM.

Another example is the channel Canal Noticias Lisboa CNL. Only in its first year of installation, it was expended about US$.10.000.000, and the equipment as well as the building were not property of the channel, but rented.

In fact, however the project will be implemented by phases, after 5 to 10 years of regular work, it is predicted to have reached a similar value in incomes as the examples above.

*value took in the Portuguese market, below the average of the European prices in about 30%.

The first ten months of works will be oriented to:
-        design of the equipment and software structures
-        start the first productions
-        elaboration of the projects of architecture and engineering

-        procurement and acquisition of all equipment
-        construction and supervision of the works
-        installation and configuration of the equipment

So, the first year must be used to the implantation of the project. However, as to accelerate the
chronogram, some operations should be started
before this timing, temporarily installed in a different building.

The first phase will be oriented to:
-        video productions (external and internal)
-        edition
-        elaboration of cost programmes

3.3   Description of the requirements of Power,
Fuel and Water. POWER
It is assumed that the unit shall use one thousand
units of power of everyday. Thus, the unit shall use
365.000 units of power in a year. The costs of power
is assumed at 3.5 cents for one unit. The consumption of power shall increase by 10% every year.

FUEL

The unit shall use D.G. set for generation of electricity, when there is an interruption in the supply. The hours of disruption in the normal supply can not be predicted accurately. Hence, the hours for which the D.G. sets shall be used also can not be predicted accurately. However, it is assumed that the unit shall utilise one thousand litters of diesel per month. The cost is assumed that 23 cents for one litter. Hence, the total cost for one thousand litters shall be US$.230. The consumption of diesel shall increase by 10% every year.

WATER

The unit does not need water for fixed commercial operation. However, it shall have a strength of 50 people as employees. Water shall be required for drinking and sanitation purposes. It is assumed that 3000 litters of water shall be required everyday. The cost of obtaining this water is assumed at US$.7 everyday. The consumption of water shall increase by 10% every year.

3.4   Description of the manpower requirement of the project and description of the steps taken by the company to train the manpower.

The total staff strength shall be 17 people in the first year.

The system of modern TV Station is designed to achieve sustained operating efficiency and transmission. This, of course, entails a certain degree of sophistication in the production and transmission. However, considering the socio-economic situation in Cape Verde, a reasonable balance has to be struck to obtain optimum performance and at the same time create gainful employment. While working out the manpower requirement for this project to be kept on direct rolls of the company totalling 50 staff and operators, the above consideration have been kept in mind.

Manpower requirement

The direct manpower required for the proposed Unit is about US$.1.260.000 per year. The manpower requirement as indicated in this chapter has been planned keeping in view the following guidelines:

Effective co-ordination among the various departments. Judicious distribution of responsibilities. Capacity utilisation of the TV Station with optimum manpower.

Details of manpower requirement for the TV Station is given in the financial section.

In line with the prevailing practice, the Security

guards, office peons and unskilled labour etc., are normally employed on contractual/casual basis, and their cost has to be included in the Factory over-heads. However in this report provision has been made to employ the above staff and their salaried have been included.

The various departments proposed shall be under the direct responsibility of the Station Director. He shall be assisted by Deputy Directors, who will look after the complete TV Station and its various day to day activities and Marketing. They shall be respon-sible for achieving the envisaged targets and sales forecasts. They shall be assisted by a team of Man-agers from production, marketing and accounts.

To run this project a labour force which shall usu-ally be composed of unskilled and skilled workers shall be employed. The first are those who do not undergo any king of specific training or education, while the latter have to do so in order to master their jobs.

When evaluating an investment project from an em-ployment point of view, its impact on both unskilled and skilled labour has been taken into account. Not only direct employment, but also indirect employ-ment has been considered. Direct employment re-fers to the new employment opportunities created within the project; indirect employment concerns job opportunities created in other projects linked with the project which is being formulated.

The implementation of large and sophisticated projects generally contributes to the development of local skills and capabilities in a country. Furthermore, they help to change traditional values, attitudes and the behaviour of the society, to build up an enterprising spirit among the people, to develop a desire for changing and improving the existing conditions of life, to introduce better work discipline and thus to change the very pattern and basis of economic development. The TV industry is already well established in Africa. Location of TV industry activities in Africa, has certain favourable factors and advantages, in setting up this type of industry, as availability of acceptable level of education among the supervisory staff adequate technical and managerial skills developed over a long period, and availability of cheap labour are assured.

The organisation structure will vary from TV Station to TV Station in the industry and as such the pattern proposed herein can be considered only as suggestive and provisional.

A well-knit organisation structure headed by a Station Director and Manager, with the supporting staff will be developed progressively during project implementation. Soon after the plant becomes operative, a good number of project staff will be absorbed in the organisation. Certain additional staff also get added to ensure smooth and efficient management of the operating unit.

The Manager Production will have a degree in communication and will be accountable and answerable to the Station Director for all TV operations including planning, production, material management, TV utilities, quality control, production cost and budgetary control, TV safety, discipline and layout relations. His main function should be to ensure achievement for quality and quantity targets of production at reasonable cost and should constantly strive to improve TV performance. In the discharge of his multifarious duties and responsibilities he will be assisted by supervisors and adequate staff for day to day activities.

His duties are to ensure that targets are maintained through effective utilisation of 3 M's viz: Machinery, Men and Materials. These targets should be translated in terms of targets for individuals under him such as supervisors, skilled workers etc. He has to ensure that the equipment at his disposal gets prompt attention on breakdown is adhered to strictly. He has to further ensure that there is a strict quality control exercised over raw materials also. He will be assisted by Deputy Manager etc. who and will report to the General Manager.

GENERAL MANAGER (Comm.) He will report to the Station Director. His main duties will be prepare sales forecasts and budgets, study and monitor the export market for the company's products and advice on ways and means of increasing sales. He will also ensure that consumer complains are solved in an appropriate manner. He will also ensure that the right materials are available for production at the right time.

He will be responsible for all aspects connected with the export procedures and keep the management updated on all matters relating to the Govt. policies on polymers and Exports and above all the world market.

COMPANY SECRETARY & FINANCIAL CONTROLLER. He will report to the Station Director and will be responsible for the departments of administration, financial planning, budgetary control, cost accounting, tax management payroll accounting, etc. and all personnel matters. He will be assisted in their duties by their respective assistants to assist in day to day activities.

Manpower planning and production

The central issue here may be one of scale. Production is staffed bye personnel, but the process may be labour-intensive or capital-intensive. Either way, planning will include:

1.   Analysis of labour supply and demand factors in relation to skills and training needs.
2.   Procedure for manpower recruitment, together with selection processes.
3.   The formation of industrial relations policies necessary for effective work place bargaining, disciplinary measures and dismissal procedures.
4.   Analysis of the effective use of human resources.
5.   Conditions necessary to maintain adequate levels of motivation. The scale of the problem is likely to be directly proportional to the method of production.

The availability of main persons is not going to be easy since TV industry is currently in its infancy in Africa.

Training needs

The selection and training of the required manpower for the proposed project has to be planned in advance.

The key personnel should be selected and trained suitably. The training would be carried out in the following manner:

Basic training on the concept of TV industry before

construction begins with visits to similar TV Stations in other countries. On site training during the construction phase of the project. On job training during the commissioning phase of the project. On job training during operation of the TV immediately after commissioning. The training of the key personnel such as Station Director should be carried out in all the phases. The training of other operating personnel should be suitably carried out during the construction, erection and operation phases in addition to training them by visits to similar plants operating in other countries.

Besides training the key operating staff described above, in TV training should also be given to other employees at skilled operating level to enable them to understand the process equipment in the project and prepare them to operate a maintain their respective sections safely, efficiently and skilfully. The above training should be carried out during construction, commissioning and operating phases of the project.

Training is necessary in order to enable personnel to acquire the skills and knowledge necessary to perform a task to an acceptable standard. The length of the training period and training methods will, of course, vary from job to job. Training is essentially a learning process, and in order that progress can be successfully monitored certain conditions are necessary.

1.   The training needs of both the individual and the organisation shall be identified and analysed.

2.   Targets and standards shall be set for the trainee, which are within his capabilities.

3.   The pace of the training programme should reflect the trainee's ability to maintain progress in properly absorbing the same.

4.   The trainee shall receive regular feedback of results. Any problem areas shall be highlighted, discussed and resolved.

5.   As the trainee progresses the amount of information provided shall be gradually reduced, thus inducing a feeling of independence and competence.

It is common place to find a wide variety of tasks in an organisation and each will require varying degrees of skill, effort and responsibility. This being so, it is inevitable that rates of pay will also vary and the differentials between the jobs will reflect their relative values. However, other factors such as local market conditions, bargaining strengths and traditions also influence a company's payment structure and a great deal of planning is required if rationalisation is to be achieved. One technique which has been successfully adopted by many companies to establish an equitable wage structure is job evaluation. In a job evaluation exercise a comparison is made of common criteria over a range of jobs, and the resulting analysis may be linked to a points allocation or job ranking system, and hence to a wage scale.

In conducting a job evaluation exercise it is important to cover a reasonable variety of tasks within the whole spectrum. For each, a job description is prepared setting out details of the duties and responsibilities undertaken by the employee together with a statement about his working conditions. Very often this task is undertaken by work study personnel since, they are responsible for determining methods of operation and evaluating the work content of the job. Each job is assessed factor by factor, resulting in a comprehensive comparative analysis.

The individual is the most important resources of any company and only people who are well trained, well motivated and adequately rewarded will provide a positive and synergistic contribution towards the company's objective and its organisational health.

In most cases, the factors, which may be weighted according to relative value, are as follows:

1.   Skill-education, experience and training.
2.   Effort-both physical and mental.

3.　　　Responsibility-for equipment, materials, initiative etc.
4.　　　Working conditions-general conditions, risk of accident and injury.

Pay policies affect not only individual employees but the whole organisation, and the rewards and objectives vary at different levels within the enterprise.

Industrial relations

An effective industrial relations policy is important, since is the system through which employees take part in decision-making and in many instances it affects the while atmosphere of employer/employee relationships. An industrial relations policy is essentially a set of rules whose determine procedures for negotiation on such matters as:

1.　　　Wage and salary scales.
2.　　　Terms and conditions of work.
3.　　　Disputes and grievances.
4.　　　Recruitment and dismissal.
5.　　　Other issues of mutual interest, e.g. closed shop, redundancies and joint consultation.

In order to promote an atmosphere of co-operation, and to minimise conflict, the needs of management and work people must be recognised by both sides. Trade unions exit to protect the interests of their members and improve their working conditions. Management, while aware of the pressures and constraints imposed by the trade unions, have a duty to maximise the use of resources at their disposal, which may be expressed in relation to profitability, return on investment, level of service, sales volume, market share and cost-effectiveness. The strategies adopted in attempting to solve industrial relations problems will vary from company to company, and indeed from union to union, but there is no doubt that they will be influenced by both internal and external factors.

Internal factors

1.	The attitudes of employees to management, and management to employees.
2.	The leadership style of management.
3.	The bargaining strength of both parties.
4.	The number of negotiating bodies.
5.	The prosperity of the company.

External factors

1.	The extent to which parent boards influence company management, and district officials influence or control local shop stewards.
2.	Whether or not bargaining is conducted at plant, local or national level.

3.   Government policy towards industrial relations.

4.   The economic situations nationally, locally
or within the company itself.

3.5   Describe the project implementation schedule.

Implementation of this project is a challenging task
and calls for meticulous planning, scheduling and
monitoring to realise the project goals within the
budgeted cost and time frame. The goal can be
achieved by adopting modern project management
techniques.

To implement this project adequately, a team of
engineers and project personnel having requisite
education and experience are being appointed, to
whom a detailed Work Breakdown Structure (WBS)
in a logical order of activities, shall be supplied
shortly, keeping in view cost estimation, scheduling,
and to help monitor and control of the project. It is
proposed to be formulated in conjunction with the
objectives of each activity and goal settings. Project
shall be programmed and controlled by network
analysis techniques. Before the application of the
network analysis techniques, the project personnel
shall be acquainted with their capabilities in saving
time, resources and costs. The training of project
personnel and engineers at levels shall be provided

for proper control of project progress and taking of timely corrective actions to re-align these efforts to meet ore stated objectives. It is proposed to gear programming and control system, i.e. project implementation system, which will ensure an integrated approach to project implementation. Project management activities shall be determined in advance and all activities carried out be project personnel as well as those to be contracted shall be identified.

Responsibility for project implementation shall be clearly defined. The forms of project organisation range from project oriented to functional organisation, while most of the cases are combinations of the two, with certain adaptations to prevailing conditions. It is impossible to over emphasise the importance of establishing a team of a task force for implementing the project with a designated leader to co-ordinate and guide its functions.

Project manager: Shall be responsible to the Board of Directors. The project manager shall be responsible for guiding and co-ordinating the efforts of all parties engaged in implementing the project, obtaining necessary government approvals on contracts. He is to control the project organisation with the promoters as well as with other agencies and organisations interested in the project. The manager shall have some staff to assist him, especially in checking expenditures to date and determining the present and future cost overrun or under run so that the project manager can take or propose to the Board pertinent corrective measures.

The network shall cover the pre-construction phase of the project indicating major administrative processes, since experience shows that some of them have frequently involved lengthy delays. In other words, it shall include the aggregate activities to be carried out be the principal parties participating in the implementation process.

The project budget shall then be prepared. Part or periodic payments to contractors which might be made at the end of certain intervals (e.g. weekly or monthly) throughout the time horizon of project implementation, shall be made by summing up activity costs per unit of time, which may be a week & month, and computing the cumulative cost at the end of each time interval. For the activities that are in process and are contacted or sub-contracted, the assumption of a linear time cost activity relationship shall be used for the sake of simplicity. In other words, expenditures are uniformly distributed throughout the duration of the activity.

PROJECT SCHEDULE

After an investment decision is taken, the main machinery and long delivery items must be ordered out at the earliest, forming the first major step in implementation of the project. It is foreseen that an engineering consultant will be appointed for carrying out the detailed engineering including basic engineering and procurement assistance to the client. It is also assumed that reputed and experienced contractors with adequate resources viz., men, materials, tools and tackles etc. will be engaged for execution of the construction and erection work. The purchase packages for auxiliaries shall be kept minimum so as to reduce the co-ordination efforts to the minimum. A great deal of co-ordination is required for constructing/erecting the new units. This task is feasible, provided the major activities of the project are co-ordinated and completed in the duration specified to achieve the respective milestones in time.

STRATEGY FOR TIMELY EXECUTION

It is important to deploy a team of experienced personnel for project execution and select the external agencies with due care for rendering the services and supply of equipment for the project. The project activities must be identified, planned and scheduled, and the progress monitored for timely project implementation. All the inputs to the project including financial resources must be identified and their inflow planned and arranged in time.

The project must be managed professionally with necessary co-ordination among the various agencies and requisite decisions taken promptly.

Establishment of an effective monitoring procedure for progress review and co-ordination.

However this tentative schedule, many activities will begin in the 6th month after the start of the financing, using different locations and training some partial personal. This will turn possible the strength of the dynamics in the whole process as well as the shorting of the estimated timing.

In short, the following key factors would constitute the broad strategy for timely execution of all activities in a pre-determined manner as per schedule shown in the bar chart as to reach at a basis of regular production.

Early selection of an effective in-house technical team (TASK FORCE) by Government of Cape Verde, headed by a Project Manager for planning and executing the project.

i)   Proper choice of external agencies such as consultants for Project Engineering., Machinery suppliers, Construction Agencies etc. keeping in view their reputation/past performance and working experience in their fields.

ii)   Adequate use of computer-based PERT/CPM techniques for project planning, scheduling and monitoring.

3.6   Description of the initial project Flowchart.

The project implementation phase embraces the period from the decision to start the project to the beginning of the commercial production. It includes a number of stages including negotiations and contracting, project design, construction and start-up.

3.7   Description of the requirement of Land.

For this project the Government of Cape Verde will participate with about 25.000 square meters of land. In selecting the land the following criteria should be followed.

1.   The land should be near to the main city of Praia.

2.   The land should be properly connected by road.

3.   It should also have proximity to many end user clients.

4.   It should also meet various Government policy of

achieving the social objectives.

2.08 Description of the requirement of Building.

The total requirement of building is as follows.
a. Adm. Building   700 Square Meters
b. Studios –3 (Three)   1.800 Square Meters
c. Other Misc. Building   300 Square Meters

Total Square Meters: approximately 2800 Square Meters.
Cost for construction of about US$750 per square meter (including air conditioning systems, primary electrical cabin etc.).

2.9   Description of some requirements for Plant & Machinery and their basis of selection.

The plant and machinery have been selected giving due consideration to the sophisticated nature of technology required. Detailed discussions were carried out with the foreign suppliers to ensure that the required capacities are practical with minimum capital and operating costs before the machinery were finally selected.

The machinery shall be selected from the most reputed foreign manufacturers of complete range of equipment. Keeping in view the following main factors:

a)   Past performance
b)   Existing machinery in Africa & abroad.
c)   Sales and service facilities in Africa

The main equipment is as follows.

STUDIO

1.   Post-production equipment – non-linear post-production, two ES7 stations.
2.   ENG – 3 DVCAM Camcorders
3.   Régie with 4 DXC-D30PK1 Digital Cameras - for cold light installation.
4.   Flexicart.
5.   Computer graphics suite.
6.   Sound room.
7.   MUST SYSTEM: Automatic Emission Managing System, with graphical information for no active emissions. (This system has been used by several French thematic channels)

Other technical data:

The whole installation is characterised by:

1.   Informational web with UTP level, 5 cables.
2.   Phone system.
3.   Specific furniture.
4.   Parking for about 200 places

Discrimination of Technical Data:

Studio

Cold lighting system.

4 Camera channels composed by:

2 DXC-D30PK Sony Digital Cameras

4 J18 Canon Zoom Lens

4   CA-537P Sony Camera Adapters
4   DXF-50 Sony Studio Viewfinder's
4   CCU-M3P Sony Control Camera Units
4   RM-7P Sony remote CCU controls
4   50M Sony Multicore Camera Sets
4   Canon Focus and Zoom control sets
4   Intercommunication Headphones
4   Special tripods

1 charriot plane and curve

1 Teleprompter with 2 reading systems (monitors

for individual speakers) 1 Lighting control table with memory – ARRI
Different microphones with diverse directional characteristics Audio components
Video components
General Intercommunication system

Video Régie

1 Recorder/Player BETACAM equipment
1 video digital mixer with 12 tracks
1 DVE equipment of digital effects for 3 channels
1 TBC Frame Synchroniser
2 Oscilloscope and Vectorscopy for control of cameras
2 Synchroniser Generators with Changeover
1 Intercommunication system with 4 places
1 Audio and Video Matrix with remote controls Video distributors

Audio Régie

1 audio mixer with 24 tracks
2 audio monitors
1 cd player
1 cassette deck
1   DAT
1   Mini-Disc
1   Level Detector for Audio Stereo Stereo Audio Distributors

Audio compressors Audio effects Microphones

Video Post-Production
3 Sony ES-7 on DVCAM/AVID hybrid non linear edition stations 1 Conventional BETACAM Edition Suite 2:1
Multi-track audio (8) on hard disc, Sound Scape type

3 ENG SETS:
Sony DSR-300 PK DVCAM Camcorder Digital Compact Report projectors
Various support materials (Batteries etc.)

3.10   Description of the Water Requirements for the project.

The requirements of water, separately for various matters are given in the following table.

Circulating	: Nil
Make-up	: Nil
Process	: Nil
Drinking	: 3,000 LPD

3.11 Description of the Steam requirement for the project.

A. Steam requirements and steam balance : N.A.

B. Capacity and type of boiler with detailed specifications : N.A.

C. Steam and energy diagram : N.A.

D.   Total energy generated / purchased (converted into M. K. : N.A. Cal) theoretical requirement of energy (in M. K. Cal)
at the various consumption stations and expected actual requirement at these stations.

E. If alternate processed are available, comparative energy : N.A.
consumption figures for the various processes. If the
project is energy intensive, possibility of choosing alternate process in order to make the project less energy
intensive.

F. Steps proposed to be taken by the company to improve : N.A.
energy losses efficiency and reduce energy losses (such
as power factor improvement, power load manage-

ment,
optimising, illumination waste heat utilisation,
etc.)

G. Scope for usage of solar / other renewable sources of   : N.A.
energy.

H. Any other measures contemplated in the direction of : N.A.
energy conservation and management.

3.123 Some information on Compressed air, fuel,
etc.

Compressed Air

(a) Requirement   : N.A.
(b) Sources   : N.A.
(c) Arrangements proposed   : N.A.
(d) Cost at site with detailed calculations : N.A.

3.13   Details of the nature of atmospheric, soil and water pollution likely to be created by the project and the measures proposed for control of pollution. Indicate whether necessary permissions for the disposal of effluent have been obtained.

There shall not be any atmospheric pollution likely to be created by the project as there is no machine which has combustion resulting into air pollution or any chemical process which may release any gases which may result into an air pollution.

However, the use of DG set shall cause a small amount of air pollution. Considering the size of DG set, the pollution is within the permissible limits.

CAPE VERDE TELEVISION NETWORK

THE INDUSTRY

CHAPTER - 4

THE INDUSTRY

4.01 Description of the Television Industry in general.

The television industry can be described in the following broad categories. 1. Introduction

2.   Cable Networks

3.   DTH

4.   TV-media

5.   Earnings drivers

6.   Outlook

The detailed discussion in each of the categories is given below.

Introduction

The growing popularity of TV as a communication

medium has resulted in the TV media sector undergoing a rapid transformation. From the black and white days of state controlled TV Station, to the highly colourful tunes of Channel V and MTV, the medium has certainly undergone a phenomenal change. Given its popularity, percentage ad spend has also increased proportionately on this medium.

Media pie (%)

	1995	1997
TV	62.5	68.8
Radio	20.9	15
Press	16.6	16.2

Source NRS

Entry of new channels post 1991

All over the world the telecasting has witnessed entry of new channels to cater to the various needs of world audiences. Channels have been launched in English as well as other regional languages. In many countries of the world till 1991, the state owned TV Station ruled the roost, as other players were not allowed to uplink and broadcast. However channels such as CNN, Star TV and BBC, which were offshore companies, could circumvent these regulations and telecast their programs into any country of the world. Cable operators then relayed the same and made it available to the common man through the cable television network.

Like many other countries, the State machinery controlled television. It was used as a propaganda tool for the party in power, with the opposition always at the receiving end. The customer had very little choice. The first steps towards more user choice began during the 1980s, which had to be telecast to a wider audience. TV Station used satellite channels for the telecast and the T.V. network was launched as an international channel.

The sports telecast by Channel 9 in 1985 and the Gulf War in the late eighties all played small but important cameos in educating the international viewer. With liberalisation in 1992 and crumbling tariff barriers televisions (read as colour TVs) became more easily available. The media revolution had started.

Major satellite channels avidly watched by viewers are Star TV, Sony TV, Home TV, BBC & CNN. There are other regional language channels which are major players in their respective territories.

Most of the channels that could not attain popularity rapidly suffered, since their advertisement earnings were not sustainable. The first round of the media wars is over. Management changes, i.e. original promoters selling out to new management with deeper pockets, has become the order of the day. Alliances like the famous ESPN Star Sports arrangement also made headlines. Given the global trends of mergers and acquisitions, further consolidation is likely. Alliances and mergers make sense when the partners complement each other, like BBC and Discovery launched Animal Planet, CNBC and ABNI came together to launch a business channel called CNBC Asia.

Cable Networks

Antennas set up by either the end user or the cable operator receives the signals transmitted by the satellite. Local cable operators lay their own cables, set up control rooms, which can telecast 40 or more channels over a limited area. They charge the household a one- time connection charge of about US$.10 per point and a recurring monthly charge ranging between US$.1 to US$.5.

Initially, this was done in a very unorganised manner. The business required local knowledge and contacts, so every locality had its own cable operator. Collection was critical for the cable operator. For the end user, quality of telecast and a complete lack of standards became an issue. This lead to a shakeout and the formation of cable companies with money power which in turn tied-up with the local and small cable operators. Cable companies charge about US$1 per month to the local cable operators and support them with training and other infrastructure inputs. The business is immensely capital intensive and takes a long time to break even.

In many countries the operations of cable operators are regulated under the Cable TV Act which ensures that pornographic materials and other materials which are against culture and values or are detrimental to national interests do not get telecast. Recently this act has been amended to include foreign channels also.

Direct To Home

DTH is a new technology that circumvents the cable operators by directly delivering a bundle of channels to the end user. DTH involves transmission of encoded audio/ video signals (Ku band) via satellite. The end user needs an antenna to receive the signals and a decoder (set top box) to unscramble the encrypted signals. DTH services elsewhere in the world are Echostar and DirecTV (USA) and BskyB (Europe). Rupert Murdoch of Star TV fame owns BskyB.

The size of the antenna in DTH will be 1.5-2 ft in diameter, making it easy to install and transport. In conventional cable, since signals are in C band, an 8ft- diameter antenna is needed. The basic difference in the business model is the hardware costs in DTH. In a cable system, the user pays a one time connect fee and monthly rentals, while in DTH he has to invest in hardware.

The antenna will cost about US$200-300 and decoder will cost about US$200. The African viewer might be reluctant to incur such heavy installation costs. Quality of telecast in DTH is superior to Cable TV and viewer can receive up to 200 channels.

DTH will result in restructuring of the cable television industry. It will become imperative to have cash reserves to withstand the technology threat. Up gradation to fibre optic backbone will become necessary. A fibre optic network will cost about US$0.5mn per km as compared to US$0.1mn per km for coaxial cable. The stage is now ripe for consolidation.

TV Media features

In the Broadcasting business, it is only the industry leader who makes sizeable profits. The business is a game of asymmetrical payoffs. For instance, the top 5 channels account for 90% of ad spend.

Urbanisation and TV penetration is related. This may be due to the popularity of cable television that has resulted in increased colour TV sales. Rural penetration is low, although growing at a fast pace, because of dearth of specific program content to cater to that segment.

Liberalisation has resulted in the world viewer becoming more aware and conscious. This has resulted in the customer having more choice with the entry of a number of companies in different segments. Competition has resulted in companies increasing their marketing spend significantly.

Popularity of TV media is becoming higher. Increasing TV penetration leads to a reallocation of advertisement budgets with higher allocation for television at the cost of other medium.

TV channel operators use different business models to generate revenues. The critical component of any channel is the quality and type of programs they telecast. This determines their popularity, which in turn determines amount of advertisement revenues they can generate. They can do any one of the following:

Buy programming rights of program software from outside and collect advertisement revenue on their own. This model is followed by several TV companies, wherein they have a separate company in their fold, which develops all the content. The advantage is that re runs of serials/ programs become very profitable.

Selling time space to the producers for a fixed charge. Producers in turn are free to book advertisements at their own rates (there is an understanding on the time allocated for advertisement) and collect revenue. This is the basic model for many TV companies in which they sell prime time slots. The rights continue to be vested with the producer.

Earnings drives

The key factors that drive sector revenues are

Television penetration: Since the medium is television, increased television penetration will imply higher viewer ship. This will translate into higher advertisement spend allocation. This will also imply higher software production and demand for new programs.

Competition from other satellite channels would have an adverse impact on advertisement revenues, as advertisers have more choice in allocating ad budgets.

Government policies can have a big impact on the fortunes of the entire industry. When the DTH bill is passed in any country then, it will trigger a restructuring of the cable business.

Launching new channels targeted at specific segments, like regional channels within any country other areas having large pockets of ethnic population would lead to revenue growth. This will entail significant initial outlays.

Depreciation of the local currency would increase revenues as most of the program/ software companies export the programs overseas and payments are dollar denominated.

Advertisement revenue

As mentioned earlier, this is the primary source of income for TV channel operators. This revenue is directly co-related with the reach and viewer ship of a channel. Any channel's popularity depends on good quality programs, which is the software content. The business requires enormous initial investment in programs and revenues follow only with a time lag after the channel receives a minimum viewer acceptance.

Outlook

The sector has latent potential for growth on back of the exponential growth of cable TVs during the last 5 years. Television penetration in Africa is extremely low as compared to other developing countries like Malaysia, Pakistan, etc. in Asia. The number of channels has increased, implying higher demand for software programs.

Advertisement revenues, which are the barometer of channel popularity, will get dispersed over several competing channels. A shakeout is likely in both the channel and cable TV sectors. The biggest beneficiaries will be the content providers or the software houses. They will control the intellectual rights to the key element driving any channel's popularity.

Direct -to-Home, Digital Terrestrial Transmission and Conditional Access Cable Delivery have emerged as new delivery mechanisms. Breakthrough in technology would help open up avenues for these channels.

CAPE VERDE TELEVISION NETWORK

MARKETING

CHAPTER - 7

MARKETING

7.1   Description of the commercial viability of the project with regards to revenue generations.

Estimations shows that, after about 5 years of regular working, the CPLP Cultural Television Channel will be able to start exporting a great quantity of services and videos to Africa, Europe and South America. (See financial projections)

The economic feasibility of the project will occur with the commercialisation of several products, like:

1.　　　The sale and production of cultural programmes to various institutions like:
- BBC
- TV Culture Brazil
- RTP
- Discovery

Etc.

All over the world commercial networks for television cultural programmes have been created. Several television channels in different countries have regularly acquired programmes by producers spread out all over the world. Not only, several international organisations, like the World Bank and the United Nations, including FAO and Unesco, among many others, need programmes for public education, like programmes oriented to alert and to educate people concerning endemic and epidemic diseases. AIDS, malaria, tuberculosis are a very few examples.

These world institutions have made a great effort to develop television programmes with humanitarian objectives. But, in Africa – undoubtedly the continent most needed of such programmes – there is no television station, in present times, with capability to make face to such a need.

Therefore, the television teams that had been responsible for such programmes are, practically in its totality, placed in countries of the called First

World, strange to the local population's concrete reality.

Programmes focusing new agricultural techniques or even oriented to agricultural, commercial and industrial education are essential elements of such a repressed demand.

The price for each institutional campaign, with an average of 10 films produced per campaign, is of about USD.15.000$ and the capacity of the CPLP Cultural Channel will be as follows:

First Group* - from 2 to 4 campaigns per month.
Second Group – from 1 campaign in two months to 1 campaign per
month

This represents:
First Group   a potential annual income of about US$.540.000
Second Group a potential annual income of about US$.90.000

*First Group: Angola, Cape Verde, and Mozambique. Second Group: Guinea-Bissau, Sao Tome e Principe, East Timor (Lorosae). Reference Group: Brazil, Portugal.

Obs. In the Second Group, for the first six years of operation it is estimated a difference of potential annual income between the Angola unit and Cape Verde's and Mozambique's respective units, from US$.540.000 to about US$.378.000.

2. The sale and production of no cultural programmes to other television channels, like:
-        series (novels)
-        talk-shows Etc.

Many countries of the region need to produce television programmes, but do not have capability to do it. Thus, they seem themselves obliged to search expensive productions in Europe. With only one programme sold-three daily hours-the channels of the First Group* would receive in incomes the equivalent of about two times of the whole investment. The Second Group countries will not have conditions at the beginning to product cultural programmes to other television channels.

*First Group: Angola, Cape Verde, and Mozambique. Second Grope: Guinea-Bissau, Sao Tome e Principe, East Timor (Lorosae). Reference Group: Brazil, Portugal.

The price for this type of television programme is:

Auditory programs   US$.12.500 per programme
Talk-shows   US$.4.000 per programme

Interviews   US$.4.000 per programme

The capacity of production, of these programmes by the countries of the First Group will be:

In the FIRST PHASE:
Talk-shows   2 per month
Interviews   4 per month
In the FINAL PHASE:
Auditory programmes   6 per month
Talk-shows   20 per month
Interviews   20 per month

Thus, the potential income will be:

FIRST PHASE
Talk-shows   US$.96.000 per year
Interviews   US$.192.000 per year

FINAL PHASE
Auditory programmes   US$.900.000 per year*
Talk-shows   US$.960.000 per year*
Interviews   US$.960.000 per year*

* the same difference as showed above.

The price of the novels is much higher. For each complete novel, with about 120 chapters, the price is about US$.2.500.000 and the capacity of the First Group of the CPLP Cultural Channel (in the final phase) will be of one novel per year.

3. There is the possibility of commercial use of the transmission time beyond the six hours reserved to culture and education.

The conventional day in television is of 18 hours, of which only 6 hours would be no commercial. We would have, therefore, 12 hours of no cultural transmission, which should be freely commercialised.
Even the period of six hours- divided into two sections of three hours, the first one dedicated to culture and second section of three hours to education-has a great potential for sponsoring.

Each commercial hour of transmission can include up to 12 minutes of advertising. The price for each 30 seconds of advertising is:

Noble time   from 6PM to 11PM   a b o u t US$.400 each 30"
Normal time   rest   about US$.200 each 30"

Concentrating the commercial programming in the noble time. With 4 hours of transmission in this period, we would have:

Educational television: from 7AM to 9AM
From 3PM to 4PM

Cultural television: from 4PM to 7PM
Commercial television: from 9AM to 3PM
From 7PM to 1AM

Thus, the commercial period would comprehend 6 hours in normal time, 4 hours in noble time and 2 hours in normal time again.

Begin 12 minutes per hour for advertising, the commercial period could have:

Noble time   48 minutes   24   films   US$.9.600 per day

Normal time   96 minutes   48   films   US$.9.600 per day

Total of the potential income............   US$.19.200 per day

> or US$.576.000 per month
> or US$.6.912.000 per year

It is believed that in the first phase of the project (after one year), the CPLP Cultural Television Channels of the First Group will be able to start with an income from commercial advertisement of about US$.1.000.000 per year-value which is predicted to increase in the follow months.

4. The rent of the studios to television teams of other countries can be another source of incomes. Many producers who cover events in Africa need to move many times to their countries of origin dur-

ing the video works, because there is not technical support in Africa. The same phenomenon happens with the cinematography and the journalism productions. It is not difficult to imagine, for example, the serious problems journalism teams have had, for example, with essential components like batteries, lighting, electronic components etc., which only can be easily find in Europe.

The period of renting of a television studio is of 12 hours. Each period has price of about US.4.000. After the first year, the studio should be rented for 12 hours per each period of 3 days. So, the rent per month, in this period (First Phase), will be able to generate incomes of about US$.120.000 per month in the First Group.

5. Support services to other television networks in all areas, including novels, mini series, docdramas series etc.

Many countries, principally in Africa, do not have technical conditions to develop this kind of programmes, but they have a strong internal repressed demand in this sense.

Each team, abroad, has a price of US$.1.000 per day (two people), after the costs of dislocation, hotel and meals. In the first months (first phase) it will be organised one team for works in other countries. In the final phase it is predicted to have up to 5 teams with such function. The capacity of each team is of about 20 days per month. Therefore, the support to other television networks-when took in its full potentiality-will be able to represent up to 20 days per month in the first phase. This represents a potential income of about US$.240.000 per year, and 100 days per month of works in the final phase, signifying about US$.1.200.000 per year of incomes, always referring to the First Group countries.

6. Colloquies, seminars, videoconferences and meetings of different natures.

The building of the CPLP Cultural Television Network will have a medium size auditorium, (First Group), with all conditions to receive seminars, colloquies and meetings of the most diverse nature. Such seminars, videoconferences etc. are important not only for the increase of the incomes, but also to attract specialists of the most different areas, being an important element for the development of all region as well as for the diffusion of the local, regional and continental cultural values.

The meetings, seminars and colloquies attract, in average, about 300 people per event. The price-excepting meals, hotels and transportation-per each participant is of about US$.50 per day. It will be a capacity for up to four events of this type per month, what could represent an income of about US$.60.000 per month or US$.720.000 per year (final phase-after 5 to 7 years).

The project should also turn possible.
-       classes of professional formation in diverse disciplines
-       support for a multimedia high technology centre (First Group)
-       support for a multimedia high cultural cen-tre (Final Group)

CAPE VERDE TELEVISION NETWORK

PROFITABILITY & CASH FLOW

CHAPTER - 8

PROFITABILITY AND CASH FLOW

8.1   Estimation of cost of production and work-ing results for the first five years of operation.

The estimated cost of production of working results for the first five years of operation are given in the chapter "Financial Projection".

8.2   Cash flow statement for the company as a whole, for five operating years of the project based on the estimates of working results.

A detailed cash flow statement for the company as a whole for five operating years is given in the chapter of "Financial Projection".

8.3   Projected balance sheet for five operating years for the company as a whole.

The balance sheet for five operation years for the company as a whole is given in the chapter of "Financial Projection".

CAPE VERDE TELEVISION NETWORK

ASSUMPTIONS

CHAPTER - 9

ASSUMPTIONS

1.   The Cape Verde Television Network shall generate the income from the following sources.

a)   Sale of cultural programmes.
b)   Sale of commercial programmes.
c)   Sale of advertisement time during transmission of TV programmes
d)   Hiring of studios.
e)   Supply of technical services
f)   Hiring of conference hall

The detailed assumptions for each of the above mentioned activities are as follows:

a)   Sale of cultural programs:

I.   It is assumed that four cultural programs shall be produced per month in the first year, which can be sold to other countries. This
figure will increase to 5 programs per month in the second year, six programs per month in the third year and so on.
II.   It is assumed that the selling price shall be US$.105 per program.

b)   Sale of commercial programs:

I.   The commercial programs consist of talk

shows, interviews and auditory programs.

II.	It is assumed that two talk shows shall be produced per month in the first year, three talk shows shall be produced in the third year, four talk shows per month shall be produced in the third year and
so on.

III.	It is assumed that the selling price of one talk show program shall
be US$.2800.

IV.	It is assumed that four interviews shall be produced per month in the first year, five interviews shall be produced in the third year, six interviews per month shall be produced in the third year and so on.

V.	It is assumed that the selling price of one interviews program shall be US$.2800.

VI.	It is assumed that no auditory programs shall be produced in the first and the second year. Only when the people have two years of experience then in the third year one auditory program shall be produced per month. In the fourth year two programs shall be produced per month and so on.

VII.	It is assumed that the selling price of one auditory program shall be UD$.8750.

c) Sale of advertisement time during the transmission of T.V. programs:

I.	There shall be a total transmission of 8 hours

per day in the first year. It will go on increasing by two hours in the second and subsequent years.

II.   Out of which advertisement shall be available for programs with transmission period of four hours in the first year. The programs in

which advertisements could be available shall increase by two hours per year in the second and subsequent years.

III.   In one hour T.V. transmission, 12 minutes of advertisement shall be allowed.

IV.   Out of total T.V. time 33 % time shall be considered as prime time and remaining 67% shall be considered as non prime time, in the first, second and third years. From the fourth year onwards the prime-time advertisements shall remain constant.

V.   The advertisement rates shall be US$ 560 for one minute of prime time advertisement and US$ 245 for one minute of non prime time advertisement.

d)   Hiring of studios:

I.   It is assumed that the studios shall be taken on hire for a shift of 12 hours, two times in a month in the first year. It will increase to

three times in a month in the second year.

II.   The rent per day shall be US$2800.

e)   Supply of technical services:-

I.   It is assumed that one team, consisting of

two technically qualified people shall be available in the first year. In the second year two such teams shall be available. In the third year three such teams shall be available and so on.

II.　　It is assumed that in the first year the team shall be hired for 10 days in a month. In the second year the teams shall be hired for 12

days in a month. In the third year the teams shall be hired for 14 days in a month and so on.

III.　　The rate of hire shall be US$.700 per team per day.

f)　　Hiring of conference hall :-

I.　　It is assumed that the conference hall shall be taken on hire for two times in a month in the first year, three times in a month in the

second year and four times in a month in the third year and so on.

II.　　The hire charges shall be US$.350 per conference.

2.　　The unit is planning transmission of three hours of educational programs, three hours of cultural programs and two hours of commercial programs per day in the first year. Thus, there shall be a total transmission of eight hours of programs per day in the first year. The total transmission hours shall increase by two hours per day in the second and subsequent years. Thus, in the second year there shall be total transmission of 10 hours per day and in the third year there shall be a total transmission of 12 hours per day. The cost of software production is assumed at US$.1750 per hour.

3.   The cost of miscellaneous consumable items is assumed at about US$.25200 in the first year. This will increase by 10% in the second and subsequent years.

4.   The cost of stores and spares consumed is assumed at US$.29400 in the first year. It will increase by 10% in the second and subsequent years.

5.   The cost of repairs and maintenance is assumed at 0.1% of the total cost of plant and machinery in the first year. This cost will increase up to 0,3% in the fifth year.

6.   It is assumed that the unit shall use 1.000 units of power every day or 365.000 units of power in the first year. The consumption of power shall increase by 10% in the second and subsequent years. The cost of power is assumed at 3.5 cents for one unit.

7.   It is assumed that the unit shall consume 700 litters of diesel per month. The cost is assumed at 33 cents for one litter. It will increase by 10% in second and subsequent years.

8.   The unit shall have a technical staff of 35 people and administrative staff of 15 people.

9.   Depreciation is calculated on reducing balance method. The rate of depreciation is taken at 10% for building, 25% for plant and machinery, 10% for furniture and fixtures and 20% on vehicles.

10.   Advertisement expenditure is assumed at 0.2% of the sales.

11.   It is assumed that the stock of raw material and stores shall be 30 days of consumption, receivables shall be 45 days sales and creditors shall be 15 days purchases.

12.   The term loan shall be for a total period of 25 years. It shall have a moratorium of five years. It shall be repayable in 20 equal yearly instalments.

PROJECT REPORT

OF

GUINEA BISSAU CULTURAL TELEVISION NETWORK

AT

GUINEA BISSAU

AFRICA

PROMOTERS

REPUBLIC OF GUINEA BISSAU

AFRICA

GUINEA BISSAU TELEVISION NETWORK

INTRODUCTION

CHAPTER - 1

INTRODUCTION

1.1   Description of the organisation through which this project is being taken up.

The CPLP is an international organisation, created in 17 of July of 1996, with Headquarters in Lisbon and consisting of the following State Members: Angola, Brazil, Cape-Verde, Guinea-Bissau, Mozambique, Portugal, Sao Tome e Principe and East Timor (Timor Lorosae).

The CPLP has as its objective the politic-diplomatic relationships between its Members, mainly referring to the United Nations and the World Bank, the relationship EU/MERCOSUL and the accomplishment of the Europe-Africa Summit.

The CPLP is also dedicated to the co-operation, particularly in the economic, social, cultural, legal and technical-scientific matters; as well as to the projects of promotion and broadcasting of the Portuguese Language, nominated to the improvement of the International Institute of the Portuguese Language and the creation of a Bibliographical Fund.

All its decisions are taken by consensus.

2.2   Description of the main reasons for setting up an African Cultural Television Network.

Africa is a continent submerged into dramatic problems. All these humanitarian problems have in the education and culture their most direct and objective root.

The African countries speaking Portuguese are among the African countries with the lowest level of development.

The whole situation is so dramatic that the implementation of the CPLP Cultural Television Network is more than urgent.

It is hoped that with this Television Network the people shall have better understanding of various diseases and other problems of life. e.g. AIDS is prevalent disease in Africa. But as the people are illiterate the information about the disease cannot be spread by any other method. In such cases the Television Network becomes a very effective audio-visual means of communication.

The project also intends to be a base of a massive

production of health campaigns, specially to reinforce the combat against the AIDS, the infect-contagious diseases- endemic or epidemic-through a process of global education of the local and regional populations.

The CPLP Cultural Television Network is a project for peace and its objective is not to compete with local television channels already installed, but to interchange with them, promoting a future global web dedicated to knowledge and development.

1.03 Description of the main objects of the project.

The major objective of the project is to assist the CPLP countries to develop the first African television network channel oriented to cultural and educational themes with the follow orientation:

-        diffusion of the Portuguese language
-        education in all areas
-        reinforcement of the local and regional cultures
-        formation of technical personal on communication and tele-education

-        formation and information for a better participation of the CPLP nations in the process of global development
-        to support the implantation of the Institute

of Portuguese Language

The development of such a project will create an irradiation pole from the countries with the project implementation to the whole African continent, constituting an African network of research and culture, discovering and valorising different cultural aspects and, consequently, reinforcing identity factors.

With the stronger cultural identity, it is expected to have lower levels of violence in social behaviour as well as a relevant increase of the educational performance in all its sectors.

Systematic educational objectives are complementary to the general orientation of this project, which is cultural in its fundament.
A very large positive impact is also to be expected with respect to several sectors of the economy, greatly improving the overall quality of services.

1.4   Description of the mission status of the project.

The mission of the project is:

-       to rescue lost values and cultural information

- to reinforce historical elements
- to rescue old artworks
- to give a new value to new artworks
- to communicate practical and objective information on health
- to communicate practical and objective information on agriculture
- to communicate practical and objective information on economy

1.05 Description of the philosophy of the project.

The CPLP Cultural Television Network will have in its philosophy, as a support to its mission, the follow fundaments:

- to work with local people
- be focused on quality
- intercommunication

Three words can define the CPLP Cultural Television Network philosophy:

People, quality and intercommunication

Following these three key words of its philosophy, a single sentence also defines its mission: "A Continent of Culture".

Two main sectors characterise its global structure:
1. Cultural Sector

1.1.   To develop programmes on:
-       Music
-       Architecture (also vernacular)
-       Archaeology
-       Literature and poetry
-       History
-       Tourism
-       Dance
-       Plastic arts
-       Astronomy
-       Science in general
-       gastronomy
-       health
-       anthropology
-       local cultures
-       others

1.2   The programmes can be of all types, the most important is to be attentive to communication. All programmes must reach the highest possible number of spectators.

2. Educational sector.
2.1   Systematic education in short practical courses.
2.2   Educational series with practical and objective information on:
-       health care
-       family
-       agriculture
-       managing

- services
- civil construction

As the CPLP Cultural Television Network must not be supported by the respective governments in the future, it will have a daily time of free commercial programmes up to 12 hours. This is a good solution to pay its costs and support its development.

1.06 Description the magnitude of the project.

The project proposes a Television transmission time of eight hours per day in the first year. It will increase by two hours per day in the second and subsequent years.

The project also envisages training of local people in all the related areas of Television. It is proposed to develop a local pool of talent which can be utilised not only in this particular country but also in other countries of Africa.

The total cost of the project of the Guinea Bissau Cultural Television Network is US$.10.600.000. Out of this it is proposed to obtain a soft loan of US$.8.800.000. The remaining amount of US$.1.800.000 shall be contributed by the Government of Guinea Bissau in terrain for the installation of the respective building, infrastructures and logistic support.

For the purpose of the Guinea Bissau Cultural Television Network a company shall be formed with the participation of the Government. The Contribution of the Government in the form of share capital shall be US$.1.800.000.

The term loan shall be on soft terms. The total period of the loan shall be 23 years. Out of which five years shall be a period of moratorium. The entire loan shall be repaid in the remaining 17 years, in 17 equal yearly instalments.

The Government of Guinea Bissau will give the sovereign guarantee for the referred loan.

GUINEA BISSAU TELEVISION NETWORK

THE COUNTRY

CHAPTER - 2

THE COUNTRY

2.1 - Description the level of development of Guinea Bissau.

One of the 20 poorest countries in the world, Guinea-Bissau depends mainly on farming and fishing.

Cashew crops have increased remarkably in the recent years, and the country now ranks sixth in cashew production.

Guinea- Bissau exports fish and seafood along with small amounts of peanuts, palm kernels, and timber. Rice is the major crop and staple food.

Trade reform and price liberalisation are the most successful part of the country's structural adjustment program under IMF sponsorship. The tightening of monetary policy and the development of the private sector have begun to reinvigorate the economy. Inflation dropped sharply in the first quarter of 1997. Membership in the WAMU (West African Monetary Union), begun in May 1997, should help support 5% annual growth and contribute to fiscal discipline. Because of high costs, the development of petroleum, phosphate, and other mineral resources is not a near-term prospect.

OFFICIAL NAME: Republic of Guinea-Bissau
CAPITAL: Bissau
SYSTEM OF GOVERNMENT: Multiparty Republic
AREA: 36,125 Sq Km (13,948 Sq Mi)
ESTIMATED 2000 POPULATION: 1,179,800

LOCATION & GEOGRAPHY: Guinea-Bissau is located on the coast of West Africa. It is bound by Guinea to the south and east, Senegal to the north and the Atlantic Ocean to the south- west. The territory consists of the mainland, the Bissagos archipelago that is a group of over 18 islands and various other coastal islands. The mainland terrain has a forested coastal plain typified by mangrove lined estuaries as well as a transitional savannah covered plateau which forms the Planalto de Bafata and Planalto de Gabu. The country is drained by a number of rivers that flow into the Atlantic Ocean and the principal rivers are the Cacheu or Farim, Mansoa, Geba, Corubal, Rio Grande and Cacine. Major Cities (pop. est.); Bissau 125,000, Bafata 14,000, Gabu 8,000 (1988). Land Use; forested 38%, pastures 38%, agricultural-cultivated 12%, other 12% (1993).

CLIMATE: Guinea-Bissau has a tropical climate with two seasons, a wet season from June to November with August the wettest month and a dry season from December to May with April and May the hottest months. The prevailing wind is the hot dust laden Harmattan which blows from the Sahara Desert in the west. Average annual precipitation in Bissau is 1,950 mm (77 inches) while average temperature ranges are from 24 degrees Celsius (75 degrees Fahrenheit) to 27 degrees Celsius (81 degrees Fahrenheit) all year.

PEOPLE: The majority of the population is of Black African origin and includes the following ethnic groups, the Balante who account for 27% while the Fulani account for 23%, the Malinke for 12%, the Mandyako for 11% and the Pepel for 10%. The Mestizos, Mulattoes and Assimilados who are of mixed racial descent, constitute the most important ethnic minorities representing around 2% of the population.

DEMOGRAPHIC/VITAL STATISTICS: Density; 27 persons per sq km (71 persons per sq mi) (1991). Urban-Rural; 19.9% urban, 80.1% rural (1990). Sex Distribution; 49.2% male, 50.8% female (1990). Life Expectancy at Birth; 39.9 years male, 43.1 years female (1990). Age Breakdown; 41% under 15, 25% 15 to 29, 17% 30 to 44, 11% 45 to 59, 5% 60 to 74, 1% 75 and over (1990). Birth Rate; 42.9 per 1,000 (1990). Death Rate; 23.0 per 1,000 (1990) . Increase

Rate; 19.9 per 1,000 (1990). Infant Mortality Rate; 151.0 per 1,000 live births (1990).

RELIGIONS: Over 65% of the population follow local native tribal beliefs while 30% are Muslims and the remainder are Christians, predominantly Roman Catholic.

LANGUAGES: The official language is Portuguese, although a Creole Patois is the trade and national language. The indigenous people all speak dialects derived from the Niger-Congo family of languages.

EDUCATION: Aged 25 or over and having attained: N/A. Literacy; literate population aged 15 or over 211,200 or 36.5% (1990).

CURRENCY: The official currency is the Peso (PG) divided into 100 Centavos.

ECONOMY: Gross National Product; USD $241,700,000(1993).PublicDebt;USD$633,600,000 (1993). Imports; USD $90,000,000 (1991). Exports; USD $23,000,000 (1991). Tourism Receipts; N/A. Balance of Trade; PG -293,146,000,000 (1994). Economically Active Population; 461,000 or 45.8% of total population (1992). Unemployed; N/A.

MAIN TRADING PARTNERS: Its main trading partners are Portugal, Egypt, Senegal, France, Spain,

Cape Verde, Germany and Algeria.

MAIN PRIMARY PRODUCTS: Bauxite, Cereals, Coconuts, Cotton, Fish, Palm Kernels, Peanuts, Rice,
Timber.

MAJOR INDUSTRIES: Agriculture, Beverages, Construction, Fishing, Food

Processing, Forestry.

MAIN EXPORTS: Fish, Ground Nuts, Palm Kernels,
Timber.

TRANSPORT: Railroads; nil. Roads; length 3,500 km
(2,175 mi) (1989). Vehicles; cars 3,200 (1989), trucks
and buses 2,400 (1989). Merchant Marine; vessels
18 (1990), dead weight tonnage 1,846 (1990). Air
Transport; passenger-km 9,000,000 (5,592,000
passenger-mi) (1985), cargo ton-km 1,000,000
(685,000 short ton-mi) (1985).

COMMUNICATIONS: Daily Newspapers; total of 1
with a total circulation of 6,000 (1992). Radio; receivers 40,000 (1994). Television; N/A. Telephones;
units 8,600 (1993).

MILITARY: 7,250 (1995) total active duty personnel with 93.8% army, 4.8% navy and 1.4% air force
while military expenditure accounts for 3.3% (1992)
of the Gross National Product (GNP).

GUINEA BISSAU TELEVISION NETWORK

THE PROJECT

CHAPTER - 3

THE PROJECT

3.1   Description of the initial steps to be taken for the implementation of the project.

The first objective of the CPLP Cultural Television Network is to install its structures in the following countries:

-        Sao Tome e Principe
-        Cape Verde
-        Mozambique
-        Guinea-Bissau
-        Angola
-        Timor

Thus, these six countries will structure the basic network of the whole project.

Angola, Mozambique and Cape Verde will form a FIRST GROUP; Sao Tome e Principe, Guinea-Bissau and East Timor will form a SECOND GROUP; and, finally, Brazil and Portugal will form a called REFERENCE GROUP.

However, the project will start at the same time in all countries. A more precise description of the meaning of each group is made below.

The basic physical components of the project are:

1.   The creation of companies, in each of the States of the CPLP, oriented to the objectives above referred.

2.   The State of each country – who will be the responsible and guarantee of the payment of the financing – will also have a participation in the companies.

3.   All strategies will be oriented to lowest costs in long term.

4.   The whole project will be developed by phases.

5.   The timetable of transmission will be defined case by case.

6.   The project will include:

6.1.   a complete plan for an equipment network.

6.2.   a complete plan for a software network.

6.3.   all procurements

6.4.   equipment transportation and installation

6.5.   all configurations

6.6.   creation of the infographic system

6.7.   technical formation of the personal
6.8.   architecture and engineer projects for the building
6.9.   construction of the building organised in phases

1.2   Description of the procedure by which the necessary programmes will be produced for transmission on the channel.

The television programs are known as software in this field. The unit is planning transmission of three hours of educational programs, three hours of cultural programs and two hours of commercial programs per day in the first year. Thus, there shall be a total transmission of eight hours of programs per day in the first year. The total transmission hours shall increase by two hours per day in the second and subsequent years. Thus, in the second year there shall be total transmission of 10 hours per day and in the third year there shall be a total transmission of 12 hours per day. The cost of software production is assumed at US$.2.500 per hour. The amount of US$.2.500 per hour is taken based on cost of production of such programmes in Portugal.

To contract a similar system from other television channel, only for one country, for a transmission period of 6 hours daily, the cost in, market (eg. Portugal*) is of about US$.500.000 per month. This cost represents a total of about US$.6.000.000 per year. In 16 years, excluding the first year of implementation, the costs for contract the six daily hours of transmission from other television channel, in present market price, is of about US$.100.000.000. The six countries would have an estimated cost of about US$.600.000.000 in a 16 years period of time.

These costs are based on the prices presently used in Portugal, as for example the programme Hora Viva – Seguranca em Directo, transmitted every day, excluding weekends, from 7 to 10 AM.

Another example is the channel Canal Noticias Lisboa CNL. Only in its first year of installation, it was expended about US$.10.000.000, and the equipment as well as the building were not property of the channel, but rented.

In fact, however the project will be implemented by phases, after 5 to 10 years of regular work, it is predicted to have reached a similar value in incomes as the examples above.

*value took in the Portuguese market, below the average of the European prices in about 30%.

The first ten months of works will be oriented to:
-        design of the equipment and software structures

- start the first productions
- elaboration of the projects of architecture and engineering
- procurement and acquisition of all equipment
- construction and supervision of the works
- installation and configuration of the equipment

So, the first year must be used to the implantation of the project. However, as to accelerate the chronogram, some operations should be started before this timing, temporarily installed in a different building.

The first phase will be oriented to:
- video productions (external and internal)
- edition
- elaboration of cost programmes

3.3   Description of the requirements of Power, Fuel and Water. POWER
It is assumed that the unit shall use one thousand units of power of everyday. Thus, the unit shall use 365.000 units of power in a year. The costs of power is assumed at two cents for one unit. The consumption of power shall increase by 10% every year.

FUEL

The unit shall use D.G. set for generation of electricity, when there is an interruption in the supply. The hours of disruption in the normal supply can not be predicted accurately. Hence, the hours for which the D.G. sets shall be used also can not be predicted accurately. However, it is assumed that the unit shall utilise one thousand litters of diesel per month. The cost is assumed in about 13 cents for one litter. Hence, the total cost for one thousand litters shall be US$.130. The consumption of diesel shall increase by 10% every year.

WATER

The unit does not need water for fixed commercial operation. However, it shall have a strength of 50 people as employees. Water shall be required for drinking and sanitation purposes. It is assumed that 3000 litters of water shall be required everyday. The cost of obtaining this water is assumed at US$.4 everyday. The consumption of water shall increase by 10% every year.

3.4   Description of the manpower requirement of the project and description of the steps taken by the company to train the manpower.

The total staff strength shall be 17 people in the first year.

The system of modern TV Station is designed to achieve sustained operating efficiency and transmission. This, of course, entails a certain degree of sophistication in the production and transmission. However, considering the socio-economic situation in Guinea Bissau, a reasonable balance has to be struck to obtain optimum performance and at the same time create gainful employment. While working out the manpower requirement for this project to be kept on direct rolls of the company totalling 50 staff and operators, the above consideration have been kept in mind.

Manpower requirement

The direct manpower required for the proposed Unit is about US$.720.000 per year. The manpower requirement as indicated in this chapter has been planned keeping in view the following guide-lines:

Effective co-ordination among the various departments. Judicious distribution of responsibilities.

Capacity utilisation of the TV Station with optimum manpower.

Details of manpower requirement for the TV Station is given in the financial section.

In line with the prevailing practice, the Security guards, office peons and unskilled labour etc., are normally employed on contractual/casual basis, and their cost has to be included in the Factory overheads. However in this report provision has been made to employ the above staff and their salaried have been included.

The various departments proposed shall be under the direct responsibility of the Station Director. He shall be assisted by Deputy Directors, who will look after the complete TV Station and its various day to day activities and Marketing. They shall be responsible for achieving the envisaged targets and sales forecasts. They shall be assisted by a team of Managers from production, marketing and accounts.

To run this project a labour force which shall usually be composed of unskilled and skilled workers shall be employed. The first are those who do not undergo any king of specific training or education, while the latter have to do so in order to master their jobs.

When evaluating an investment project from an employment point of view, its impact on both unskilled and skilled labour has been taken into account. Not only direct employment, but also indirect employment has been considered. Direct employment refers to the new employment opportunities created within the project; indirect employment concerns job opportunities created in other projects linked with the project which is being formulated.

The implementation of large and sophisticated projects generally contributes to the development of local skills and capabilities in a country. Furthermore, they help to change traditional values, attitudes and the behaviour of the society, to build up an enterprising spirit among the people, to develop a desire for changing and improving the existing conditions of life, to introduce better work discipline and thus to change the very pattern and basis of economic development. The TV industry is already well established in Africa. Location of TV industry activities in Africa, has certain favourable factors and advantages, in setting up this type of industry, as availability of acceptable level of education among the supervisory staff adequate technical and managerial skills developed over a long period, and availability of cheap labour are assured.

The organisation structure will vary from TV Station to TV Station in the industry and as such the pattern proposed herein can be considered only as suggestive and provisional.

A well-knit organisation structure headed by a Station Director and Manager, with the supporting staff will be developed progressively during project implementation. Soon after the plant becomes operative, a good number of project staff will be absorbed in the organisation. Certain additional staff also get added to ensure smooth and efficient management of the operating unit.

The Manager Production will have a degree in communication and will be accountable and answerable to the Station Director for all TV operations including planning, production, material management, TV utilities, quality control, production cost and budgetary control, TV safety, discipline and layout relations. His main function should be to ensure achievement for quality and quantity targets of production at reasonable cost and should constantly strive to improve TV performance. In the discharge of his multifarious duties and responsibilities he will be assisted by supervisors and adequate staff for day to day activities.

His duties are to ensure that targets are maintained through effective utilisation of 3 M's viz: Machinery, Men and Materials. These targets should be translated in terms of targets for individuals under him such as supervisors, skilled workers etc. He has to ensure that the equipment at his disposal gets prompt attention on breakdown is adhered to strictly. He has to further ensure that there is a strict quality control exercised over raw materials also. He will be assisted by Deputy Manager etc. who and will report to the General Manager.

GENERAL MANAGER (Comm.) He will report to the Station Director. His main duties will be prepare sales forecasts and budgets, study and monitor the export market for the company's products and advice on ways and means of increasing sales. He will also ensure that consumer complains are solved in an appropriate manner. He will also ensure that the right materials are available for production at the right time.

He will be responsible for all aspects connected with the export procedures and keep the management updated on all matters relating to the Govt. policies on polymers and Exports and above all the world market.

COMPANY SECRETARY & FINANCIAL CONTROLLER. He will report to the Station Director and will be responsible for the departments of administration, financial planning, budgetary control, cost accounting, tax management payroll accounting, etc. and all personnel matters. He will be assisted in their duties by their respective assistants to assist in day to day activities.

Manpower planning and production

The central issue here may be one of scale. Production is staffed bye personnel, but the process may be labour-intensive or capital-intensive. Either way, planning will include:

1.   Analysis of labour supply and demand factors in relation to skills and training needs.
2.   Procedure for manpower recruitment, together with selection processes.
3.   The formation of industrial relations policies necessary for effective work place bargaining, disciplinary measures and dismissal procedures.
4.   Analysis of the effective use of human resources.
5.   Conditions necessary to maintain adequate levels of motivation. The scale of the problem is likely to be directly proportional to the method of production.

The availability of main persons is not going to be easy since TV industry is currently in its infancy in Africa.

Training needs

The selection and training of the required manpower for the proposed project has to be planned in advance.

The key personnel should be selected and trained suitably. The training would be carried out in the following manner:

Basic training on the concept of TV industry before construction begins with visits to similar TV Stations in other countries. On site training during the construction phase of the project. On job training during the commissioning phase of the project. On job training during operation of the TV immediately after commissioning. The training of the key personnel such as Station Director should be carried out in all the phases. The training of other operating personnel should be suitably carried out during the construction, erection and operation phases in addition to training them by visits to similar plants operating in other countries.

Besides training the key operating staff described above, in TV training should also be given to other employees at skilled operating level to enable them to understand the process equipment in the project and prepare them to operate a maintain their respective sections safely, efficiently and skilfully. The above training should be carried out during construction, commissioning and operating phases of the project.

Training is necessary in order to enable personnel to acquire the skills and knowledge necessary to perform a task to an acceptable standard. The length of the training period and training methods will, of course, vary from job to job. Training is essentially a learning process, and in order that progress can be successfully monitored certain conditions are necessary.

1.   The training needs of both the individual and the organisation shall be identified and analysed.

2.   Targets and standards shall be set for the trainee, which are within his capabilities.

3.   The pace of the training programme should reflect the trainee's ability to maintain progress in properly absorbing the same.

4.   The trainee shall receive regular feedback of results. Any problem areas shall be highlighted, discussed and resolved.

5.   As the trainee progresses the amount of information provided shall be gradually reduced, thus inducing a feeling of independence and competence.

It is common place to find a wide variety of tasks in an organisation and each will require varying degrees of skill, effort and responsibility. This being so, it is inevitable that rates of pay will also vary and the differentials between the jobs will reflect their relative values. However, other factors such as local market conditions, bargaining strengths and traditions also influence a company's payment structure and a great deal of planning is required if rationalisation is to be achieved. One technique which has been successfully adopted by many companies to establish an equitable wage structure is job evaluation. In a job evaluation exercise a comparison is made of common criteria over a range of jobs, and the resulting analysis may be linked to a points allocation or job ranking system, and hence to a wage scale.

In conducting a job evaluation exercise it is important to cover a reasonable variety of tasks within the whole spectrum. For each, a job description is prepared setting out details of the duties and responsibilities undertaken by the employee together with a statement about his working conditions. Very often this task is undertaken by work study personnel since, they are responsible for determining methods of operation and evaluating the work content of the job. Each job is assessed factor by factor, resulting in a comprehensive comparative analysis.

The individual is the most important resources of any company and only people who are well trained, well motivated and adequately rewarded will provide a positive and synergistic contribution towards the company's objective and its organisational health.

In most cases, the factors, which may be weighted according to relative value, are as follows:

1.   Skill-education, experience and training.

2.   Effort-both physical and mental.

3.   Responsibility-for equipment, materials, initiative etc.

4.   Working conditions-general conditions, risk of accident and injury.

Pay policies affect not only individual employees but the whole organisation, and the rewards and objectives vary at different levels within the enterprise.

Industrial relations

An effective industrial relations policy is important, since is the system through which employees take part in decision-making and in many instances it affects the while atmosphere of employer/employee relationships. An industrial relations policy is essentially a set of rules whose determine procedures for negotiation on such matters as:

1.   Wage and salary scales.

2.   Terms and conditions of work.

3.   Disputes and grievances.

4.   Recruitment and dismissal.

5.   Other issues of mutual interest, e.g. closed shop, redundancies and joint consultation.

In order to promote an atmosphere of co-operation, and to minimise conflict, the needs of management and work people must be recognised by both sides. Trade unions exit to protect the interests of their members and improve their working conditions. Management, while aware of the pressures and constraints imposed by the trade unions, have a duty to maximise the use of resources at their disposal, which may be expressed in relation to profitability, return on investment, level of service, sales volume, market share and cost-effectiveness. The strategies adopted in attempting to solve industrial relations problems will vary from company to company, and indeed from union to union, but there is no doubt that they will be influenced by both internal and external factors.

Internal factors

1.   The attitudes of employees to management, and management to employees.
2.   The leadership style of management.
3.   The bargaining strength of both parties.
4.   The number of negotiating bodies.
5.   The prosperity of the company.

External factors

1.   The extent to which parent boards influence company management, and district officials

influence or control local shop stewards.

2.   Whether or not bargaining is conducted at plant, local or national level.

3.   Government policy towards industrial relations.

4.   The economic situations nationally, locally or within the company itself.

3.5   Describe the project implementation schedule.

Implementation of this project is a challenging task and calls for meticulous planning, scheduling and monitoring to realise the project goals within the budgeted cost and time frame. The goal can be achieved by adopting modern project management techniques.

To implement this project adequately, a team of engineers and project personnel having requisite education and experience are being appointed, to whom a detailed Work Breakdown Structure (WBS) in a logical order of activities, shall be supplied shortly, keeping in view cost estimation, scheduling, and to help monitor and control of the project. It is proposed to be formulated in conjunction with the objectives of each activity and goal settings. Project shall be programmed and controlled by network analysis techniques. Before the application of the network analysis techniques, the project personnel shall be acquainted with their capabilities in saving time, resources and costs. The training of project personnel and engineers at levels shall be provided for proper control of project progress and taking of timely corrective actions to re-align these efforts to meet ore stated objectives. It is proposed to gear programming and control system, i.e. project implementation system, which will ensure an integrated approach to project implementation. Project management activities shall be determined in advance and all activities carried out be project personnel as well as those to be contracted shall be identified.

Responsibility for project implementation shall be clearly defined. The forms of project organisation range from project oriented to functional organisation, while most of the cases are combinations of the two, with certain adaptations to prevailing conditions. It is impossible to over emphasise the importance of establishing a team of a task force for implementing the project with a designated leader to co-ordinate and guide its functions.

Project manager: Shall be responsible to the Board of Directors. The project manager shall be responsible for guiding and co-ordinating the efforts of all parties engaged in implementing the project, obtaining necessary government approvals on contracts. He is to control the project organisation with the promoters as well as with other agencies and organisations interested in the project. The manager shall have some staff to assist him, especially in checking expenditures to date and determining the present and future cost overrun or under run so that the project manager can take or propose to the Board pertinent corrective measures.

The network shall cover the pre-construction phase of the project indicating major administrative processes, since experience shows that some of them have frequently involved lengthy delays. In other words, it shall include the aggregate activities to be carried out be the principal parties participating in the implementation process.

The project budget shall then be prepared. Part or periodic payments to contractors which might be made at the end of certain intervals (e.g. weekly or monthly) throughout the time horizon of project implementation, shall be made by summing up activity costs per unit of time, which may be a week & month, and computing the cumulative cost at the end of each time interval. For the activities that are in process and are contacted or sub-contracted, the assumption of a linear time cost activity relationship shall be used for the sake of simplicity. In other words, expenditures are uniformly distributed throughout the duration of the activity.

PROJECT SCHEDULE

After an investment decision is taken, the main machinery and long delivery items must be ordered out at the earliest, forming the first major step in implementation of the project. It is foreseen that an engineering consultant will be appointed for carrying out the detailed engineering including basic engineering and procurement assistance to the client. It is also assumed that reputed and experienced contractors with adequate resources viz., men, materials, tools and tackles etc. will be engaged for execution of the construction and erection work. The purchase packages for auxiliaries shall be kept minimum so as to reduce the co-ordination efforts to the minimum. A great deal of co-ordination is required for constructing/erecting the new units. This task is feasible, provided the major activities of the project are co-ordinated and completed in the duration specified to achieve the respective milestones in time.

STRATEGY FOR TIMELY EXECUTION

It is important to deploy a team of experienced personnel for project execution and select the external agencies with due care for rendering the services and supply of equipment for the project. The project activities must be identified, planned and scheduled, and the progress monitored for timely project implementation. All the inputs to the project including financial resources must be identified and their inflow planned and arranged in time.

The project must be managed professionally with necessary co-ordination among the various agencies and requisite decisions taken promptly.

Establishment of an effective monitoring procedure for progress review and co-ordination.

However this tentative schedule, many activities will begin in the 6th month after the start of the financing, using different locations and training some partial personal. This will turn possible the strength of the dynamics in the whole process as well as the shorting of the estimated timing.

In short, the following key factors would constitute the broad strategy for timely execution of all activities in a pre-determined manner as per schedule shown in the bar chart as to reach at a basis of regular production.

Early selection of an effective in-house technical team (TASK FORCE) by Government of Guinea Bissau, headed by a Project Manager for planning and executing the project.

i)   Proper choice of external agencies such as consultants for Project Engineering., Machinery suppliers, Construction Agencies etc. keeping in view their reputation/past performance and working experience in their fields.

ii)   Adequate use of computer-based PERT/ CPM techniques for project planning, scheduling and monitoring.

3.6   Description of the initial project Flowchart.

The project implementation phase embraces the period from the decision to start the project to the beginning of the commercial production. It includes a number of stages including negotiations and contracting, project design, construction and start-up.

3.7   Description of the requirement of Land.

For this project the Government of Guinea Bissau will participate with about 25.000 square meters of land. In selecting the land the following criteria should be followed.

1.   The land should be near to the main city of Bissau.

2.   The land should be properly connected by road.

3.   It should also have proximity to many end user clients.

4.   It should also meet various Government policy of
        achieving the social objectives.

2.08 Description of the requirement of Building.

The total requirement of building is as follows.
a. Adm. Building   400 Square Meters
b. Studios –2 (Two)   1000 Square Meters
c. Other Misc. Building   200 Square Meters

Total Square Meters: approximately 1600 Square Meters.
Cost for construction of about US$.750 per square meter (including air conditioning systems, primary electrical cabin etc.).

2.9   Description of some requirements for Plant & Machinery and their basis of selection.

The plant and machinery have been selected giving due consideration to the sophisticated nature of technology required. Detailed discussions were carried out with the foreign suppliers to ensure that the required capacities are practical with minimum capital and operating costs before the machinery were finally selected.

The machinery shall be selected from the most reputed foreign manufacturers of complete range of equipment. Keeping in view the following main factors:

a)   Past performance
b)   Existing machinery in Africa & abroad.
c)   Sales and service facilities in Africa

The main equipment is as follows.

STUDIO

1.   Post-production equipment – non-linear post-production, two ES7 stations.
2.   ENG – 2 DVCAM Camcorders
3.   Régie with 2 DXC-D30PK1 Digital Cameras - for cold light installation.

4.   Flexicart.
5.   Computer graphics suite.

Other technical data:

The whole installation is characterised by:
1. Informational web with UTP level, 5 cables.
2.   Phone system.
3.   Specific furniture.

Discrimination of Technical Data:

Studio
Cold lighting system.
2 Camera channels composed by:
1   DXC-D30PK Sony Digital Cameras 2 J18 Canon Zoom Lens

2 CA-537P Sony Camera Adapters
1   DXF-50 Sony Studio Viewfinder's
1   CCU-M3P Sony Control Camera Units 2 RM-7P Sony remote CCU controls

2 50M Sony Multicore Camera Sets
2 Canon Focus and Zoom control sets
2 Intercommunication Headphones
2 Special tripods
1   charriot plane and curve

1   Teleprompter with 2 reading systems (monitors for individual speakers) Different microphones with diverse directional characteristics

Audio components Video components

Video Régie

1 Recorder/Player BETACAM equipment
1 video digital mixer with 12 tracks
1 DVE equipment of digital effects for 3 channels
1 TBC Frame Synchroniser
1 Oscilloscope and Vectorscopy for control of cameras
1 Synchroniser Generators with Changeover
1 Intercommunication system with 4 places
1 Audio and Video Matrix with remote controls Video distributors

Audio Régie

1 audio mixer with 24 tracks
1 audio monitors
1 cd player
1 cassette deck
1   DAT
1   Mini-Disc
1   Level Detector for Audio Stereo Stereo Audio Distributors

Audio compressors Audio effects Microphones

Video Post-Production
2 Sony ES-7 on DVCAM/AVID hybrid non linear edition stations 1 Conventional BETACAM Edition Suite 2:1
Multi-track audio (8) on hard disc, Sound Scape type

3 ENG SETS:
Sony DSR-300 PK DVCAM Camcorder Digital Compact Report projectors
Various support materials (Batteries etc.)

3.10   Description of the Water Requirements for the project.

The requirements of water, separately for various matters are given in the following table.

Circulating   : Nil
Make-up   : Nil
Process   : Nil
Drinking   : 3,000 LPD

3.11 Description of the Steam requirement for the project.

A. Steam requirements and steam balance   : N.A.

B. Capacity and type of boiler with detailed specifications   : N.A.

C. Steam and energy diagram : N.A.

D.   Total energy generated / purchased (converted into M. K. : N.A. Cal) theoretical requirement
of energy (in M. K. Cal)

at the various consumption stations and expected
actual requirement at these stations.

E. If alternate processed are available, comparative
energy : N.A.
consumption figures for the various processes. If
the
project is energy intensive, possibility of choosing
alternate process in order to make the project less
energy
intensive.

F. Steps proposed to be taken by the company to
improve   : N.A.
energy losses efficiency and reduce energy losses
(such
as power factor improvement, power load management,
optimising, illumination waste heat utilisation,

etc.)

G. Scope for usage of solar / other renewable sources of   : N.A.
energy.

H. Any other measures contemplated in the direction of : N.A.
energy conservation and management.

3.123 Some information on Compressed air, fuel,
etc.

Compressed Air

(a) Requirement   : N.A.
(b) Sources   : N.A.
(c) Arrangements proposed   : N.A.
(d) Cost at site with detailed calculations : N.A.

3.13   Details of the nature of atmospheric, soil
and water pollution likely to be created by the
project and the measures proposed for control of
pollution. Indicate whether necessary permissions
for the disposal of effluent have been obtained.

There shall not be any atmospheric pollution likely
to be created by the project as there is no machine
which has combustion resulting into air pollution or
any chemical process which may release any gases
which may result into an air pollution.

However, the use of DG set shall cause a small amount of air pollution. Considering the size of DG set, the pollution is within the permissible limits.

GUINEA BISSAU TELEVISION NETWORK

THE INDUSTRY

CHAPTER - 4

THE INDUSTRY

4.01 Description of the Television Industry in general.

The television industry can be described in the following broad categories. 1. Introduction
2.   Cable Networks

3.   DTH

4.   TV-media

5.   Earnings drivers

6.   Outlook

The detailed discussion in each of the categories is given below.

Introduction

The growing popularity of TV as a communication medium has resulted in the TV media sector undergoing a rapid transformation. From the black and white days of state controlled TV Station, to the highly colourful tunes of Channel V and MTV, the medium has certainly undergone a phenomenal change. Given its popularity, percentage ad spend has also increased proportionately on this medium.

Media pie (%)

	1995	1997
TV	62.5	68.8
Radio	20.9	15
Press	16.6	16.2

Source NRS

Entry of new channels post 1991

All over the world the telecasting has witnessed entry of new channels to cater to the various needs of world audiences. Channels have been launched in English as well as other regional languages. In many countries of the world till 1991, the state owned TV Station ruled the roost, as other players were not allowed to uplink and broadcast. However channels such as CNN, Star TV and BBC, which were offshore companies, could circumvent these regulations and telecast their programs into any country of the world. Cable operators then relayed the same and made it available to the common man through the cable television network.

Like many other countries, the State machinery controlled television. It was used as a propaganda tool for the party in power, with the opposition always at the receiving end. The customer had very little choice. The first steps towards more user choice began during the 1980s, which had to be telecast to a wider audience. TV Station used satellite channels for the telecast and the T.V. network was launched as an international channel.

The sports telecast by Channel 9 in 1985 and the Gulf War in the late eighties all played small but important cameos in educating the international viewer. With liberalisation in 1992 and crumbling tariff barriers televisions (read as colour TVs) became more easily available. The media revolution had started.

Major satellite channels avidly watched by viewers are Star TV, Sony TV, Home TV, BBC & CNN. There are other regional language channels which are major players in their respective territories.

Most of the channels that could not attain popularity rapidly suffered, since their advertisement earnings were not sustainable. The first round of the media wars is over. Management changes, i.e. original promoters selling out to new management with deeper pockets, has become the order of the day. Alliances like the famous ESPN Star Sports arrangement also made headlines. Given the global trends of mergers and acquisitions, further consolidation is likely. Alliances and mergers make sense when the partners complement each other, like BBC and Discovery launched Animal Planet, CNBC and ABNI came together to launch a business channel called CNBC Asia.

Cable Networks

Antennas set up by either the end user or the cable operator receives the signals transmitted by the satellite. Local cable operators lay their own cables, set up control rooms, which can telecast 40 or more channels over a limited area. They charge the household a one- time connection charge of about US$.10 per point and a recurring monthly charge ranging between US$.1 to US$.5.

Initially, this was done in a very unorganised manner. The business required local knowledge and contacts, so every locality had its own cable operator. Collection was critical for the cable operator. For the end user, quality of telecast and a complete lack of standards became an issue. This lead to a shakeout and the formation of cable companies with money power which in turn tied-up with the local and small cable operators. Cable companies charge about US$1 per month to the local cable operators and support them with training and other infrastructure inputs. The business is immensely capital intensive and takes a long time to break even.

In many countries the operations of cable operators are regulated under the Cable TV Act which ensures that pornographic materials and other materials which are against culture and values or are detrimental to national interests do not get telecast. Recently this act has been amended to include foreign channels also.

Direct To Home

DTH is a new technology that circumvents the cable operators by directly delivering a bundle of channels to the end user. DTH involves transmission of encoded audio/ video signals (Ku band) via satellite. The end user needs an antenna to receive the signals and a decoder (set top box) to unscramble the encrypted signals. DTH services elsewhere in the world are Echostar and DirecTV (USA) and BskyB (Europe). Rupert Murdoch of Star TV fame owns BskyB.

The size of the antenna in DTH will be 1.5-2 ft in diameter, making it easy to install and transport. In conventional cable, since signals are in C band, an 8ft- diameter antenna is needed. The basic difference in the business model is the hardware costs in DTH. In a cable system, the user pays a one time connect fee and monthly rentals, while in DTH he has to invest in hardware.

The antenna will cost about US$200-300 and decoder will cost about US$200. The African viewer might be reluctant to incur such heavy installation costs. Quality of telecast in DTH is superior to Cable TV and viewer can receive up to 200 channels.

DTH will result in restructuring of the cable television industry. It will become imperative to have cash reserves to withstand the technology threat. Up gradation to fibre optic backbone will become necessary. A fibre optic network will cost about US$0.5mn per km as compared to US$0.1mn per km for coaxial cable. The stage is now ripe for consolidation.

TV Media features

In the Broadcasting business, it is only the industry leader who makes sizeable profits. The business is a game of asymmetrical payoffs. For instance, the top 5 channels account for 90% of ad spend.

Urbanisation and TV penetration is related. This may be due to the popularity of cable television that has resulted in increased colour TV sales. Rural penetration is low, although growing at a fast pace, because of dearth of specific program content to cater to that segment.

Liberalisation has resulted in the world viewer becoming more aware and conscious. This has resulted in the customer having more choice with the entry of a number of companies in different segments. Competition has resulted in companies increasing their marketing spend significantly.

Popularity of TV media is becoming higher. Increasing TV penetration leads to a reallocation of advertisement budgets with higher allocation for television at the cost of other medium.

TV channel operators use different business models to generate revenues. The critical component of any channel is the quality and type of programs they telecast. This determines their popularity, which in turn determines amount of advertisement revenues they can generate. They can do any one of the following:

Buy programming rights of program software from outside and collect advertisement revenue on their own. This model is followed by several TV companies, wherein they have a separate company in their fold, which develops all the content. The advantage is that re runs of serials/ programs become very profitable.

Selling time space to the producers for a fixed charge. Producers in turn are free to book advertisements at their own rates (there is an understanding on the time allocated for advertisement) and collect revenue. This is the basic model for many TV companies in which they sell prime time slots. The rights continue to be vested with the producer.

Earnings drives

The key factors that drive sector revenues are

Television penetration: Since the medium is television, increased television penetration will imply higher viewer ship. This will translate into higher advertisement spend allocation. This will also imply higher software production and demand for new programs.

Competition from other satellite channels would have an adverse impact on advertisement revenues, as advertisers have more choice in allocating ad budgets.

Government policies can have a big impact on the fortunes of the entire industry. When the DTH bill is passed in any country then, it will trigger a restructuring of the cable business.

Launching new channels targeted at specific segments, like regional channels within any country other areas having large pockets of ethnic population would lead to revenue growth. This will entail significant initial outlays.

Depreciation of the local currency would increase revenues as most of the program/ software companies export the programs overseas and payments are dollar denominated.

Advertisement revenue

As mentioned earlier, this is the primary source of income for TV channel operators. This revenue is directly co-related with the reach and viewer ship of a channel. Any channel's popularity depends on good quality programs, which is the software content. The business requires enormous initial investment in programs and revenues follow only with a time lag after the channel receives a minimum viewer acceptance.

Outlook

The sector has latent potential for growth on back of the exponential growth of cable TVs during the last 5 years. Television penetration in Africa is extremely low as compared to other developing countries like Malaysia, Pakistan, etc. in Asia. The number of channels has increased, implying higher demand for software programs.

Advertisement revenues, which are the barometer of channel popularity, will get dispersed over several competing channels. A shakeout is likely in both the channel and cable TV sectors. The biggest beneficiaries will be the content providers or the software houses. They will control the intellectual rights to the key element driving any channel's popularity.

Direct-to-Home, Digital Terrestrial Transmission and Conditional Access Cable Delivery have emerged as new delivery mechanisms. Breakthrough in technology would help open up avenues for these channels.

GUINEA BISSAU TELEVISION NETWORK

MARKETING

CHAPTER - 7

MARKETING

7.1 Description of the commercial viability of the project with regards to revenue generations.

Estimations shows that, after about 5 years of regular working, the CPLP Cultural Television Channel will be able to start exporting a great quantity of services and videos to Africa, Europe and South America. (See financial projections)

The economic feasibility of the project will occur with the commercialisation of several products, like:

1.   The sale and production of cultural pro-grammes to various institutions like:
-       BBC
-       TV Culture Brazil
-       RTP
-       Discovery
Etc.

All over the world commercial networks for tel-evision cultural programmes have been created. Several television channels in different countries have regularly acquired programmes by produc-ers spread out all over the world. Not only, several international organisations, like the World Bank and the United Nations, including FAO and Unesco, among many others, need programmes for public education, like programmes oriented to alert and to educate people concerning endemic and epi-demic diseases. AIDS, malaria, tuberculosis are a very few examples.

These world institutions have made a great effort to develop television programmes with humanitar-ian objectives. But, in Africa – undoubtedly the continent most needed of such programmes – there is no television station, in present times, with capability to make face to such a need.

Therefore, the television teams that had been responsible for such programmes are, practically in its totality, placed in countries of the called First World, strange to the local population's concrete reality.

Programmes focusing new agricultural techniques or even oriented to agricultural, commercial and industrial education are essential elements of such a repressed demand.

The price for each institutional campaign, with an average of 10 films produced per campaign, is of about USD.15.000$ and the capacity of the CPLP Cultural Channel will be as follows:

First Group* - from 2 to 4 campaigns per month.
Second Group – from 1 campaign in two months to 1 campaign per month

This represents:
First Group   a potential annual income of about US$.540.000
Second Group a potential annual income of about US$.90.000

*First Group: Angola, Cape Verde, and Mozambique. Second Group: Guinea-Bissau, Sao Tome e Principe, East Timor (Lorosae). Reference Group: Brazil, Portugal.

Obs. For the case of Guinea-Bissau and East Timor, it is expected a decrease of such incomes for US$.216.000 and US$.36.000.

2. The sale and production of no cultural programmes to other television channels, like:
-        series (novels)
-        talk-shows Etc.

Many countries of the region need to produce television programmes, but do not have capability to do it. Thus, they seem themselves obliged to search expensive productions in Europe. With only one programme sold-three daily hours-the channels of the First Group* would receive in incomes the equivalent of about two times of the whole investment. The Second Group countries will not have conditions at the beginning to product cultural programmes to other television channels.

*First Group: Angola, Cape Verde, and Mozambique. Second Grope: Guinea-Bissau, Sao Tome e Principe, East Timor (Lorosae). Reference Group: Brazil, Portugal.

The price for this type of television programme is:

Auditory programs   US$.12.500 per programme
Talk-shows   US$.4.000 per programme
Interviews   US$.4.000 per programme

The capacity of production, of these programmes by the countries of the First Group will be:

In the FIRST PHASE:
Talk-shows   2 per month
Interviews   4 per month
In the FINAL PHASE:
Auditory programmes   6 per month
Talk-shows   20 per month
Interviews   20 per month

Thus, the potential income will be:

FIRST PHASE
Talk-shows   US$.96.000 per year
Interviews   US$.192.000 per year

FINAL PHASE
Auditory programmes   US$.900.000 per year
Talk-shows   US$.960.000 per year
Interviews   US$.960.000 per year

Obs. All values must be decreased for Guinea-Bissau and East Timor in the first years, as it is referred in the detailed feasibility studies.

The price of the novels is much higher. For each complete novel, with about 120 chapters, the price is about US$.2.500.000 and the capacity of the First Group of the CPLP Cultural Channel (in the final phase) will be of one novel per year.

3. There is the possibility of commercial use of the transmission time beyond the six hours reserved to culture and education.

The conventional day in television is of 18 hours, of which only 6 hours would be no commercial. We would have, therefore, 12 hours of no cultural transmission, which should be freely commercialised.
Even the period of six hours- divided into two sections of three hours, the first one dedicated to culture and second section of three hours to education-has a great potential for sponsoring.

Each commercial hour of transmission can include up to 12 minutes of advertising. The price for each 30 seconds of advertising is:

Noble time   from 6PM to 11PM   a b o u t US$.400 each 30"
Normal time   rest   about US$.200 each 30"

Concentrating the commercial programming in the noble time. With 4 hours of transmission in this period, we would have:

Educational television: from 7AM to 9AM
From 3PM to 4PM
Cultural television: from 4PM to 7PM
Commercial television: from 9AM to 3PM
From 7PM to 1AM

Thus, the commercial period would comprehend 6 hours in normal time, 4 hours in noble time and 2 hours in normal time again.

Begin 12 minutes per hour for advertising, the commercial period could have:

Noble time   48 minutes   24 films US$.9.600 per day
Normal time   96 minutes   48 films US$.9.600 per day
Total of the potential income............ US$.19.200 per day

               or US$.576.000 per month
               or US$.6.912.000 per year

It is believed that in the first phase of the project (after one year), the CPLP Cultural Television Channels of the First Group will be able to start with an income from commercial advertisement of about US$.1.000.000 per year-value which is predicted to increase in the follow months.

4. The rent of the studios to television teams of other countries can be another source of incomes. Many producers who cover events in Africa need to move many times to their countries of origin during the video works, because there is not technical support in Africa. The same phenomenon happens with the cinematography and the journalism productions. It is not difficult to imagine, for example, the serious problems journalism teams have had, for example, with essential components like batteries, lighting, electronic components etc., which only can be easily find in Europe.

The period of renting of a television studio is of 12 hours. Each period has price of about US.4.000. After the first year, the studio should be rented for 12 hours per each period of 3 days. So, the rent per month, in this period (First Phase), will be able to generate incomes of about US$.120.000 per month in the First Group.

5. Support services to other television networks in all areas, including novels, mini series, docdramas series etc.

Many countries, principally in Africa, do not have technical conditions to develop this kind of programmes, but they have a strong internal repressed demand in this sense.

Each team, abroad, has a price of US$.1.000 per day

(two people), after the costs of dislocation, hotel and meals. In the first months (first phase) it will be organised one team for works in other countries. In the final phase it is predicted to have up to 5 teams with such function. The capacity of each team is of about 20 days per month. Therefore, the support to other television networks-when took in its full potentiality-will be able to represent up to 20 days per month in the first phase. This represents a potential income of about US$.240.000 per year, and 100 days per month of works in the final phase, signifying about US$.1.200.000 per year of incomes, always referring to the First Group countries.

6. Colloquies, seminars, videoconferences and meetings of different natures.

The building of the CPLP Cultural Television Network will have a medium size auditorium, (First Group), with all conditions to receive seminars, colloquies and meetings of the most diverse nature. Such seminars, videoconferences etc. are important not only for the increase of the incomes, but also to attract specialists of the most different areas, being an important element for the development of all region as well as for the diffusion of the local, regional and continental cultural values.

The meetings, seminars and colloquies attract, in average, about 300 people per event. The price-excepting meals, hotels and transportation-per each participant is of about US$.50 per day. It will be a capacity for up to four events of this type per month, what could represent an income of about US$.60.000 per month or US$.720.000 per year (final phase-after 5 to 7 years).

The project should also turn possible.
-        classes of professional formation in diverse disciplines
-        support for a multimedia high technology centre (First Group)
-        support for a multimedia high cultural centre (Final Group)

GUINEA BISSAU TELEVISION NETWORK

PROFITABILITY & CASH FLOW

CHAPTER - 8

PROFITABILITY AND CASH FLOW

8.1   Estimation of cost of production and working results for the first five years of operation.

The estimated cost of production of working results

for the first five years of operation are given in the chapter "Financial Projection".

8.2   Cash flow statement for the company as a whole, for five operating years of the project based on the estimates of working results.

A detailed cash flow statement for the company as a whole for five operating years is given in the chapter of "Financial Projection".

8.3   Projected balance sheet for five operating years for the company as a whole.

The balance sheet for five operation years for the company as a whole is given in the chapter of "Financial Projection".

GUINEA BISSAU TELEVISION NETWORK

ASSUMPTIONS

CHAPTER - 9

ASSUMPTIONS

1.   The Guinea Bissau Television Network shall generate the income from the following sources.

a)   Sale of cultural programmes.
b)   Sale of commercial programmes.
c)   Sale of advertisement time during transmission of TV programmes

d)   Hiring of studios.
e)   Supply of technical services
f)   Hiring of conference hall

The detailed assumptions for each of the above mentioned activities are as follows:

a)   Sale of cultural programs:

I.   It is assumed that four cultural programs shall be produced per month in the first year, which can be sold to other countries. This figure will increase to 5 programs per month in the second year, six programs per month in the third year and so on.

II.   It is assumed that the selling price shall be US$.60 per program.

b)   Sale of commercial programs:

I.   The commercial programs consist of talk shows, interviews and auditory programs.

II.   It is assumed that two talk shows shall be produced per month in the first year, three talk shows shall be produced in the third year, four talk shows per month shall be produced in the third year and

so on.

III.   It is assumed that the selling price of one talk show program shall

be US$.1600.

IV.   It is assumed that four interviews shall be produced per month in the first year, five interviews shall be produced in the third year, six interviews per month shall be produced in the third year and so on.

V.   It is assumed that the selling price of one interviews program shall be US$.1600.

VI.   It is assumed that no auditory programs shall be produced in the first and the second year. Only when the people have two years of experience then in the third year one auditory program shall be produced per month. In the fourth year two programs shall be produced per month and so on.

VII.   It is assumed that the selling price of one auditory program shall be UD$.5.000.

c) Sale of advertisement time during the transmission of T.V. programs:

I.   There shall be a total transmission of 8 hours per day in the first year. It will go on increasing by two hours in the second and subsequent years.

II.   Out of which advertisement shall be available for programs with transmission period of four hours in the first year. The programs in

which advertisements could be available shall increase by two hours per year in the second and subsequent years.

III.   In one hour T.V. transmission, 12 minutes of advertisement shall be allowed.

IV.   Out of total TV time 33% time shall be considered as prime time and remaining 67% shall be considered as non prime time, in the first, second and third years. From the fourth year onwards the prime-time advertisements shall remain constant.

V.   The advertisement rates shall be US$ 320 for one minute of prime time advertisement and US$ 140 for one minute of non prime time advertisement.

d)   Hiring of studios:

I.   It is assumed that the studios shall be taken on hire for a shift of 12 hours, two times in a month in the first year. It will increase to

three times in a month in the second year.
II.   The rent per day shall be US$.1.600.

e)   Supply of technical services:-

I.   It is assumed that one team, consisting of two technically qualified people shall be available in the first year. In the second year two

such teams shall be available. In the third year three such teams shall be available and so on.
II.   It is assumed that in the first year the team shall be hired for 10 days in a month. In the second year the teams shall be hired for 12

days in a month. In the third year the teams shall be hired for 14 days in a month and so on.

III.   The rate of hire shall be US$.400 per team per day.

f)   Hiring of conference hall :-

I.   It is assumed that the conference hall shall be taken on hire for two times in a month in the first year, three times in a month in the

second year and four times in a month in the third year and so on.

II.   The hire charges shall be US$.200 per conference.

2.   The unit is planning transmission of three hours of educational programs, three hours of cultural programs and two hours of commercial programs per day in the first year. Thus, there shall be a total transmission of eight hours of programs per day in the first year. The total transmission hours shall increase by two hours per day in the second and subsequent years. Thus, in the second year there shall be total transmission of 10 hours per day and in the third year there shall be a total transmission of 12 hours per day. The cost of software production is assumed at US$.1.000 per hour.

3.   The cost of miscellaneous consumable items is assumed at US$.14.400 in the first year. This will increase by 10% in the second and subse-

quent years.

4.	The cost of stores and spares consumed is assumed at US$.16.800 in the first year. It will increase by 10% in the second and subsequent years.

5.	The cost of repairs and maintenance is assumed at 0.1% of the total cost of plant and machinery in the first year. This cost will increase up to 0,3% in the fifth year.

6.	It is assumed that the unit shall use 1.000 units of power every day or 365.000 units of power in the first year. The consumption of power shall increase by 10% in the second and subsequent years. The cost of power is assumed at 2 cents for one unit.

7.	It is assumed that the unit shall consume 1.000 litters of diesel per month. The cost is assumed at 13 cents for one litter. It will increase by 10% in second and subsequent years.

8.	The unit shall have a technical staff of 35 people and administrative staff of 15 people.

9.	Depreciation is calculated on reducing balance method. The rate of depreciation is taken at 10% for building, 25% for plant and machinery, 10% for furniture and fixtures and 20% on vehicles.

10.   Advertisement expenditure is assumed at 0.2% of the sales.

11.   It is assumed that the stock of raw material and stores shall be 30 days of consumption, receivables shall be 45 days sales and creditors shall be 15 days purchases.

12.   The term loan shall be for a total period of 25 years. It shall have a moratorium of five years. It shall be repayable in 20 equal yearly instalments.

PROJECT REPORT

OF

MOZAMBIQUE CULTURAL TELEVISION NETWORK

AT

MOZAMBIQUE

AFRICA

PROMOTERS

REPUBLIC OF MOZAMBIQUE

AFRICA

MOZAMBIQUE TELEVISION NETWORK

INTRODUCTION

CHAPTER - 1

INTRODUCTION

1.1   Description of the organisation through which this project is being taken up.

The CPLP is an international organisation, created in 17 of July of 1996, with Headquarters in Lisbon and consisting of the following State Members: Angola, Brazil, Cape-Verde, Guinea-Bissau, Mozambique, Portugal, Sao Tome e Principe and East Timor (Timor Lorosae).

The CPLP has as its objective the politic-diplomatic relationships between its Members, mainly referring to the United Nations and the World Bank, the relationship EU/MERCOSUL and the accomplishment of the Europe-Africa Summit.

The CPLP is also dedicated to the co-operation, particularly in the economic, social, cultural, legal and technical-scientific matters; as well as to the projects of promotion and broadcasting of the Por-

tuguese Language, nominated to the improvement of the International Institute of the Portuguese Language and the creation of a Bibliographical Fund.

All its decisions are taken by consensus.

2.2   Description of the main reasons for setting up an African Cultural Television Network.

Africa is a continent submerged into dramatic problems. All these humanitarian problems have in the education and culture their most direct and objective root.

The African countries speaking Portuguese are among the African countries with the lowest level of development.

The whole situation is so dramatic that the implementation of the CPLP Cultural Television Network is more than urgent.

It is hoped that with this Television Network the people shall have better understanding of various diseases and other problems of life. e.g. AIDS is prevalent disease in Africa. But as the people are illiterate the information about the disease cannot be spread by any other method. In such cases the Television Network becomes a very effective audio-visual means of communication.

The project also intends to be a base of a massive production of health campaigns, specially to reinforce the combat against the AIDS, the infect-contagious diseases- endemic or epidemic-through a process of global education of the local and regional populations.

The CPLP Cultural Television Network is a project for peace and its objective is not to compete with local television channels already installed, but to interchange with them, promoting a future global web dedicated to knowledge and development.

1.03 Description of the main objects of the project.

The major objective of the project is to assist the CPLP countries to develop the first African television network channel oriented to cultural and educational themes with the follow orientation:

-        diffusion of the Portuguese language
-        education in all areas
-        reinforcement of the local and regional cultures
-        formation of technical personal on communication and tele-education
-        formation and information for a better participation of the CPLP nations in the process of global development
-        to support the implantation of the Institute

of Portuguese Language

The development of such a project will create an irradiation pole from the countries with the project implementation to the whole African continent, constituting an African network of research and culture, discovering and valorising different cultural aspects and, consequently, reinforcing identity factors.

With the stronger cultural identity, it is expected to have lower levels of violence in social behaviour as well as a relevant increase of the educational performance in all its sectors.

Systematic educational objectives are complementary to the general orientation of this project, which is cultural in its fundament.
A very large positive impact is also to be expected with respect to several sectors of the economy, greatly improving the overall quality of services.

1.4   Description of the mission status of the project.

The mission of the project is:

-        to rescue lost values and cultural information
-        to reinforce historical elements

-       to rescue old artworks
-       to give a new value to new artworks
-       to communicate practical and objective information on health
-       to communicate practical and objective information on agriculture
-       to communicate practical and objective information on economy

1.05 Description of the philosophy of the project.

The CPLP Cultural Television Network will have in its philosophy, as a support to its mission, the follow fundaments:

-       to work with local people
-       be focused on quality
-       intercommunication

Three words can define the CPLP Cultural Television Network philosophy:
People, quality and intercommunication

Following these three key words of its philosophy, a single sentence also defines its mission: "A Continent of Culture".

Two main sectors characterise its global structure:
1. Cultural Sector
1.1.   To develop programmes on:
-       Music

- Architecture (also vernacular)
- Archaeology
- Literature and poetry
- History
- Tourism
- Dance
- Plastic arts
- Astronomy
- Science in general
- gastronomy
- health
- anthropology
- local cultures
- others

1.2   The programmes can be of all types, the most important is to be attentive to communication. All programmes must reach the highest possible number of spectators.

2. Educational sector.
2.1   Systematic education in short practical courses.
2.2   Educational series with practical and objective information on:
- health care
- family
- agriculture
- managing
- services
- civil construction

As the CPLP Cultural Television Network must not be supported by the respective governments in the future, it will have a daily time of free commercial programmes up to 12 hours. This is a good solution to pay its costs and support its development.

1.06 Description the magnitude of the project.

The project proposes a Television transmission time of eight hours per day in the first year. It will increase by two hours per day in the second and subsequent years.

The project also envisages training of local people in all the related areas of Television. It is proposed to develop a local pool of talent which can be utilised not only in this particular country but also in other countries of Africa.

The total cost of the project of the Mozambique Cultural Television Network is US$.18.400.000. Out of this it is proposed to obtain a soft loan of US$.15.400.000. The remaining amount of US$.3.000.000 shall be contributed by the Government of Mozambique in terrain for the installation of the respective building, infrastructures and logistic support.

For the purpose of the Mozambique Cultural Television Network a company shall be formed with the participation of the Government. The Contribution of the Government in the form of share capital shall be US$.3.000.000.

The term loan shall be on soft terms. The total period of the loan shall be 23 years. Out of which five years shall be a period of moratorium. The entire loan shall be repaid in the remaining 17 years, in 17 equal yearly instalments.

The Government of Mozambique will give the sovereign guarantee for the referred loan.

MOZAMBIQUE TELEVISION NETWORK

THE COUNTRY

CHAPTER - 2

THE COUNTRY

2.1 - Description of the level of development of Mozambique.

Before the peace accord of October 1992, Mozambique had been devastated by civil war and was one of the poorest countries on the globe.

Prospects subsequently improved, and with its solid economic performance in 1996-97, Mozambique has begun to exploit its agricultural, hydropower, and transportation resources. Foreign assistance programs help supply the foreign exchange required to support the budget and pay for imports of goods and services.

The restoration of electrical transmission lines to South Africa and the completion of a new transmission line to Zimbabwe (permitting the giant Cahora Bassa hydropower plant to export large amounts of electricity), proposed construction of a natural gas pipeline to South Africa, and reform of transportation services will greatly improve foreign exchange receipts. The Mozambique and South African Governments are developing the Maputo corridor, linking the port of Maputo with Witbank, South Africa. In the past few years, more than 700 state enterprises have been privatized, including the country's largest commercial bank and a number of sizeable manufacturing firms. Other pending reform measures are the reform of tax collection and the facilitation of private enterprise in the transportation, energy, and telecommunications sectors.

OFFICIAL NAME: Republic of Mozambique
CAPITAL: Maputo
SYSTEM OF GOVERNMENT: Unitary Multiparty Republic
AREA: 798,800 Sq Km (308,418 Sq Mi)
ESTIMATED 2000 POPULATION: 17,857,100

LOCATION & GEOGRAPHY: Mozambique is located on the south-east coast of Africa. It is bound by Swaziland to the south, South Africa to the south-west, Zimbabwe to the west, Zambia and Malawi to the north-west, Tanzania to the north and the Indian Ocean to the east. The country is divided into two topographical regions by the Zambezi River. (1.) North of the Zambezi river, the narrow coastline moves inland to hills and low plateaux, and further west to rugged highlands, which include the Livingstone-Nyasa Highlands, Namuli or Shire

Highlands, Angonia Highlands, Tete Highlands and the Maconde Plateau. (2.) South of the Zambezi River, the lowlands are broader with the Mashonaland Plateau and Lebomo Mountains located in the deep south. The country is drained by five principal rivers and several smaller ones with the largest and most important the Zambezi. The country has three lakes, Lake Nyasa or Malawi, Lake Chiuta and Lake Shirwa, all in the north. Major Cities (pop. est.); Maputo 931,600, Beira 298,800, Nampula 250,500 (1991). Land Use; forested 18%, pastures 56%, agricultural-cultivated 4%, other 22% (1993).

CLIMATE: Mozambique has a tropical climate with two seasons. A wet season from October to March and a dry season from April to September. Climatic conditions, however, vary depending on altitude. Rainfall is heavy along the coast and decreases in the north and south. Annual precipitation varies from 500 to 900 mm (20 to 35 inches) depending on the region with an average of 590 mm (23 inches). Cyclones are also common during the wet season. Average temperature ranges in Maputo are from 13 to 24 degrees Celsius (55 to 75 degrees Fahrenheit) in July to 22 to 31 degrees Celsius (72 to 88 degrees Fahrenheit) in February.

PEOPLE: The majority of the population belong to local tribal groups which include the Makua-Lomwe who account for 37% of the population while the Shona account for 10% and the Tsonga for 23%.

Other ethnic minorities include Europeans, mainly Portuguese, Euro-Asians and Indians.

DEMOGRAPHIC/VITAL STATISTICS: Density; 18 persons per sq km (47 persons per sq mi) (1991). Urban-Rural; 13.2% urban, 86.8% rural (1980). Sex Distribution; 49.3% male, 50.7% female (1990). Life Expectancy at Birth; 44.9 years male, 48.1 years female (1990). Age Breakdown; 44% under 15, 26% 15 to 29, 16% 30 to 44, 9% 45 to 59, 4% 60 to 74, 1% 75 and over (1990). Birth Rate; 45.0 per 1,000 (1990). Death Rate; 18.5 per 1,000 (1990). Increase Rate; 26.5 per 1,000 (1990). Infant Mortality Rate; 141.0 per 1,000 live births (1990).

RELIGIONS: Around 48% of the population follow local native tribal beliefs while 39% are Christians and 13% are Muslims.

LANGUAGES: The official language is Portuguese which is used for government, education and commerce purposes. A variety of local tribal languages are also widely spoken.

EDUCATION: Aged 25 or over and having attained: no formal schooling 80.7%, primary 18.2%, secondary 0.9%, higher 0.2% (1980). Literacy; literate population aged 15 or over 32.9% (1990).

CURRENCY: The official currency is the Metical (MT) (plural; Meticais) divided into 100 Centavos.

ECONOMY: Gross National Product; USD $1,375,000,000 (1993). Public Debt; USD $4,650,000,000 (1993). Imports; USD $1,018,000,000 (1994). Exports; USD $149,500,000 (1994). Tourism Receipts; N/A. Balance of Trade; USD - $868,000,000 (1994). Economically Active Population; 5,671,290 or 48.6% of total population (1980). Unemployed; 1.7% (1980).

MAIN TRADING PARTNERS: Its main trading partners are the USA, Portugal, France, Iraq, Japan and Singapore.

MAIN PRIMARY PRODUCTS : Bananas, Bauxite, Cashew Nuts, Cassava, Cereals, Coal, Coconuts, Copper.

MAJOR INDUSTRIES: Agriculture, Chemicals, Cement, Food Processing,

Mining, Petroleum Products, Textiles.

MAIN EXPORTS: Cashew Nuts, Cotton, Shrimp, Tea, Textiles.

TRANSPORT: Railroads; route length 3,271 km (2,033 mi) (1988), passenger-km 75,300,000 (46,789,000 passenger- mi) (1988), cargo ton- km 231,800,000 (158,760,000 short ton-mi) (1988). Roads; length 26,095 km (16,215 mi) (1989). Vehicles; cars 84,000 (1989), trucks and buses 24,000 (1989). Merchant Marine; vessels 114 (1990), dead weight tonnage 29,153 (1990). Air Transport; passenger-km 553,829,000 (344,133,000 passenger-mi) (1990), cargo ton-km 60,636,000 (41,530,000 short ton-mi) (1990).

COMMUNICATIONS: Daily Newspapers; total of 2 with a total circulation of 81,000 (1994). Radio; receivers 620,000 (1994). Television; receivers 35,000 (1994). Telephones; units 62,100 (1993).

MILITARY: 50,000 (1993) total active duty personnel with 90.0% army, 2.2% navy and 8.0% air force while military expenditure accounts for 7.6% (1993) of the Gross National Product (GNP).

MOZAMBIQUE TELEVISION NETWORK

THE PROJECT

CHAPTER - 3

THE PROJECT

3.1   Description of the initial steps to be taken for the implementation of the project.

The first objective of the CPLP Cultural Television Network is to install its structures in the following countries:

-        Sao Tome e Principe
-        Angola
-        Mozambique
-        Guinea-Bissau
-        Mozambique
-        Timor

Thus, these six countries will structure the basic network of the whole project.

Mozambique, Mozambique and Angola will form a FIRST GROUP; Sao Tome e Principe, Guinea-Bissau and East Timor will form a SECOND GROUP; and, finally, Brazil and Portugal will form a called REFERENCE GROUP.

However, the project will start at the same time in all countries. A more precise description of the meaning of each group is made below.

The basic physical components of the project are:

1.   The creation of companies, in each of the States of the CPLP, oriented to the objectives above referred.
2.   The State of each country – who will be the responsible and guarantee of the payment of the financing – will also have a participation in the companies.
3.   All strategies will be oriented to lowest costs in long term.
4.   The whole project will be developed by phases.
5.   The timetable of transmission will be defined case by case.
6.   The project will include:
6.1.   a complete plan for an equipment network.
6.2.   a complete plan for a software network.
6.3.   all procurements
6.4.   equipment transportation and installation
6.5.   all configurations
6.6.   creation of the infographic system

6.7.   technical formation of the personal

6.8.   architecture and engineer projects for the building

6.9.   construction of the building organised in phases

1.2   Description of the procedure by which the necessary programmes will be produced for transmission on the channel.

The television programs are known as software in this field. The unit is planning transmission of three hours of educational programs, three hours of cultural programs and two hours of commercial programs per day in the first year. Thus, there shall be a total transmission of eight hours of programs per day in the first year. The total transmission hours shall increase by two hours per day in the second and subsequent years. Thus, in the second year there shall be total transmission of 10 hours per day and in the third year there shall be a total transmission of 12 hours per day. The cost of software production is assumed at US$.2.500 per hour. The amount of US$.2.500 per hour is taken based on cost of production of such programmes in Portugal.

To contract a similar system from other television channel, only for one country, for a transmission period of 6 hours daily, the cost in, market (eg. Portugal*) is of about US$.500.000 per month. This cost represents a total of about US$.6.000.000 per year. In 16 years, excluding the first year of implementation, the costs for contract the six daily hours of transmission from other television channel, in present market price, is of about US$.100.000.000. The six countries would have an estimated cost of about US$.600.000.000 in a 16 years period of time.

These costs are based on the prices presently used in Portugal, as for example the programme Hora Viva – Seguranca em Directo, transmitted every day, excluding weekends, from 7 to 10 AM.

Another example is the channel Canal Noticias Lisboa CNL. Only in its first year of installation, it was expended about US$.10.000.000, and the equipment as well as the building were not property of the channel, but rented.

In fact, however the project will be implemented by phases, after 5 to 10 years of regular work, it is predicted to have reached a similar value in incomes as the examples above.

*value took in the Portuguese market, below the average of the European prices in about 30%.

The first ten months of works will be oriented to:
-        design of the equipment and software structures

-        start the first productions
-        elaboration of the projects of architecture and engineering
-        procurement and acquisition of all equipment
-        construction and supervision of the works
-        installation and configuration of the equipment

So, the first year must be used to the implantation of the project. However, as to accelerate the chronogram, some operations should be started before this timing, temporarily installed in a different building.

The first phase will be oriented to:
-        video productions (external and internal)
-        edition
-        elaboration of cost programmes

3.3   Description of the requirements of Power, Fuel and Water. POWER
It is assumed that the unit shall use one thousand units of power of everyday. Thus, the unit shall use 365.000 units of power in a year. The costs of power is assumed at 3.5 cents for one unit. The consumption of power shall increase by 10% every year.

FUEL

The unit shall use D.G. set for generation of electricity, when there is an interruption in the supply. The hours of disruption in the normal supply can not be predicted accurately. Hence, the hours for which the D.G. sets shall be used also can not be predicted accurately. However, it is assumed that the unit shall utilise one thousand litters of diesel per month. The cost is assumed that 23 cents for one litter. Hence, the total cost for one thousand litters shall be US$.230. The consumption of diesel shall increase by 10% every year.

WATER

The unit does not need water for fixed commercial operation. However, it shall have a strength of 50 people as employees. Water shall be required for drinking and sanitation purposes. It is assumed that 3000 litters of water shall be required everyday. The cost of obtaining this water is assumed at US$.7 everyday. The consumption of water shall increase by 10% every year.

3.4   Description of the manpower requirement of the project and description of the steps taken by the company to train the manpower.

The total staff strength shall be 17 people in the first year.

The system of modern TV Station is designed to achieve sustained operating efficiency and transmission. This, of course, entails a certain degree of sophistication in the production and transmission. However, considering the socio-economic situation in Mozambique, a reasonable balance has to be struck to obtain optimum performance and at the same time create gainful employment. While working out the manpower requirement for this project to be kept on direct rolls of the company totalling 50 staff and operators, the above consideration have been kept in mind.

Manpower requirement

The direct manpower required for the proposed Unit is about US$.1.260.000 per year. The manpower requirement as indicated in this chapter has been planned keeping in view the following guidelines:

Effective co-ordination among the various departments. Judicious distribution of responsibilities. Capacity utilisation of the TV Station with optimum manpower.

Details of manpower requirement for the TV Station is given in the financial section.

In line with the prevailing practice, the Security guards, office peons and unskilled labour etc., are normally employed on contractual/casual basis, and their cost has to be included in the Factory overheads. However in this report provision has been made to employ the above staff and their salaried have been included.

The various departments proposed shall be under the direct responsibility of the Station Director. He shall be assisted by Deputy Directors, who will look after the complete TV Station and its various day to day activities and Marketing. They shall be responsible for achieving the envisaged targets and sales forecasts. They shall be assisted by a team of Managers from production, marketing and accounts.

To run this project a labour force which shall usually be composed of unskilled and skilled workers shall be employed. The first are those who do not undergo any king of specific training or education, while the latter have to do so in order to master their jobs.

When evaluating an investment project from an employment point of view, its impact on both unskilled and skilled labour has been taken into account. Not only direct employment, but also indirect employment has been considered. Direct employment refers to the new employment opportunities created within the project; indirect employment concerns job opportunities created in other projects linked with the project which is being formulated.

The implementation of large and sophisticated projects generally contributes to the development of local skills and capabilities in a country. Furthermore, they help to change traditional values, attitudes and the behaviour of the society, to build up an enterprising spirit among the people, to develop a desire for changing and improving the existing conditions of life, to introduce better work discipline and thus to change the very pattern and basis of economic development. The TV industry is already well established in Africa. Location of TV industry activities in Africa, has certain favourable factors and advantages, in setting up this type of industry, as availability of acceptable level of education among the supervisory staff adequate technical and managerial skills developed over a long period, and availability of cheap labour are assured.

The organisation structure will vary from TV Station to TV Station in the industry and as such the pattern proposed herein can be considered only as suggestive and provisional.

A well-knit organisation structure headed by a Station Director and Manager, with the supporting staff will be developed progressively during project implementation. Soon after the plant becomes operative, a good number of project staff will be absorbed in the organisation. Certain additional staff also get added to ensure smooth and efficient management of the operating unit.

The Manager Production will have a degree in communication and will be accountable and answerable to the Station Director for all TV operations including planning, production, material management, TV utilities, quality control, production cost and budgetary control, TV safety, discipline and layout relations. His main function should be to ensure achievement for quality and quantity targets of production at reasonable cost and should constantly strive to improve TV performance. In the discharge of his multifarious duties and responsibilities he will be assisted by supervisors and adequate staff for day to day activities.

His duties are to ensure that targets are maintained through effective utilisation of 3 M's viz: Machinery, Men and Materials. These targets should be translated in terms of targets for individuals under him such as supervisors, skilled workers etc. He has to ensure that the equipment at his disposal gets prompt attention on breakdown is adhered to strictly. He has to further ensure that there is a strict quality control exercised over raw materials also. He will be assisted by Deputy Manager etc. who and will report to the General Manager.

GENERAL MANAGER (Comm.) He will report to the Station Director. His main duties will be prepare sales forecasts and budgets, study and monitor the export market for the company's products and advice on ways and means of increasing sales. He will also ensure that consumer complains are solved in an appropriate manner. He will also ensure that the right materials are available for production at the right time.

He will be responsible for all aspects connected with the export procedures and keep the management updated on all matters relating to the Govt. policies on polymers and Exports and above all the world market.

COMPANY SECRETARY & FINANCIAL CONTROLLER. He will report to the Station Director and will be responsible for the departments of administration, financial planning, budgetary control, cost accounting, tax management payroll accounting, etc. and all personnel matters. He will be assisted in their duties by their respective assistants to assist in day to day activities.

Manpower planning and production

The central issue here may be one of scale. Production is staffed bye personnel, but the process may be labour-intensive or capital-intensive. Either way, planning will include:

1.   Analysis of labour supply and demand factors in relation to skills and training needs.
2.   Procedure for manpower recruitment, together with selection processes.
3.   The formation of industrial relations policies necessary for effective work place bargaining, disciplinary measures and dismissal procedures.
4.   Analysis of the effective use of human resources.
5.   Conditions necessary to maintain adequate levels of motivation. The scale of the problem is likely to be directly proportional to the method of production.

The availability of main persons is not going to be easy since TV industry is currently in its infancy in Africa.

Training needs

The selection and training of the required manpower for the proposed project has to be planned in advance.

The key personnel should be selected and trained suitably. The training would be carried out in the following manner:

Basic training on the concept of TV industry before construction begins with visits to similar TV Stations in other countries. On site training during the construction phase of the project. On job training during the commissioning phase of the project. On job training during operation of the TV immediately after commissioning. The training of the key personnel such as Station Director should be carried out in all the phases. The training of other operating personnel should be suitably carried out during the construction, erection and operation phases in addition to training them by visits to similar plants operating in other countries.

Besides training the key operating staff described above, in TV training should also be given to other employees at skilled operating level to enable them to understand the process equipment in the project and prepare them to operate a maintain their respective sections safely, efficiently and skilfully. The above training should be carried out during construction, commissioning and operating phases of the project.

Training is necessary in order to enable personnel to acquire the skills and knowledge necessary to perform a task to an acceptable standard. The length of the training period and training methods will, of course, vary from job to job. Training is essentially a learning process, and in order that progress can be successfully monitored certain conditions are necessary.

1.   The training needs of both the individual and the organisation shall be identified and analysed.

2.   Targets and standards shall be set for the trainee, which are within his capabilities.

3.   The pace of the training programme should reflect the trainee's ability to maintain progress in properly absorbing the same.

4.   The trainee shall receive regular feedback of results. Any problem areas shall be highlighted, discussed and resolved.

5.   As the trainee progresses the amount of information provided shall be gradually reduced, thus inducing a feeling of independence and competence.

It is common place to find a wide variety of tasks in an organisation and each will require varying degrees of skill, effort and responsibility. This being so, it is inevitable that rates of pay will also vary and the differentials between the jobs will reflect their relative values. However, other factors such as local market conditions, bargaining strengths and traditions also influence a company's payment structure and a great deal of planning is required if rationalisation is to be achieved. One technique which has been successfully adopted by many companies to establish an equitable wage structure is job evaluation. In a job evaluation exercise a comparison is made of common criteria over a range of jobs, and the resulting analysis may be linked to a points allocation or job ranking system, and hence to a wage scale.

In conducting a job evaluation exercise it is important to cover a reasonable variety of tasks within the whole spectrum. For each, a job description is prepared setting out details of the duties and responsibilities undertaken by the employee together with a statement about his working conditions. Very often this task is undertaken by work study personnel since, they are responsible for determining methods of operation and evaluating the work content of the job. Each job is assessed factor by factor, resulting in a comprehensive comparative analysis.

The individual is the most important resources of any company and only people who are well trained, well motivated and adequately rewarded will provide a positive and synergistic contribution towards the company's objective and its organisational health.

In most cases, the factors, which may be weighted according to relative value, are as follows:

1.   Skill-education, experience and training.

2.   Effort-both physical and mental.
3.   Responsibility-for equipment, materials, initiative etc.
4.   Working conditions-general conditions, risk of accident and injury.

Pay policies affect not only individual employees but the whole organisation, and the rewards and objectives vary at different levels within the enterprise.

Industrial relations

An effective industrial relations policy is important, since is the system through which employees take part in decision-making and in many instances it affects the while atmosphere of employer/employee relationships. An industrial relations policy is essentially a set of rules whose determine procedures for negotiation on such matters as:

1.   Wage and salary scales.
2.   Terms and conditions of work.
3.   Disputes and grievances.
4.   Recruitment and dismissal.
5.   Other issues of mutual interest, e.g. closed shop, redundancies and joint consultation.

In order to promote an atmosphere of co-operation, and to minimise conflict, the needs of management and work people must be recognised by both sides. Trade unions exit to protect the interests of their members and improve their working conditions. Management, while aware of the pressures and constraints imposed by the trade unions, have a duty to maximise the use of resources at their disposal, which may be expressed in relation to profitability, return on investment, level of service, sales volume, market share and cost-effectiveness. The strategies adopted in attempting to solve industrial relations problems will vary from company to company, and indeed from union to union, but there is no doubt that they will be influenced by both internal and external factors.

Internal factors

1.   The attitudes of employees to management, and management to employees.
2.   The leadership style of management.
3.   The bargaining strength of both parties.
4.   The number of negotiating bodies.
5.   The prosperity of the company.

External factors

1.   The extent to which parent boards influence company management, and district officials influence or control local shop stewards.
2.   Whether or not bargaining is conducted at plant, local or national level.

3.   Government policy towards industrial relations.

4.   The economic situations nationally, locally or within the company itself.

3.5   Describe the project implementation schedule.

Implementation of this project is a challenging task and calls for meticulous planning, scheduling and monitoring to realise the project goals within the budgeted cost and time frame. The goal can be achieved by adopting modern project management techniques.

To implement this project adequately, a team of engineers and project personnel having requisite education and experience are being appointed, to whom a detailed Work Breakdown Structure (WBS) in a logical order of activities, shall be supplied shortly, keeping in view cost estimation, scheduling, and to help monitor and control of the project. It is proposed to be formulated in conjunction with the objectives of each activity and goal settings. Project shall be programmed and controlled by network analysis techniques. Before the application of the network analysis techniques, the project personnel shall be acquainted with their capabilities in saving time, resources and costs. The training of project personnel and engineers at levels shall be provided for proper control of project progress and taking of timely corrective actions to re-align these efforts to meet ore stated objectives. It is proposed to gear programming and control system, i.e. project implementation system, which will ensure an integrated approach to project implementation. Project management activities shall be determined in advance and all activities carried out be project personnel as well as those to be contracted shall be identified.

Responsibility for project implementation shall be clearly defined. The forms of project organisation range from project oriented to functional organisation, while most of the cases are combinations of the two, with certain adaptations to prevailing conditions. It is impossible to over emphasise the importance of establishing a team of a task force for implementing the project with a designated leader to co-ordinate and guide its functions.

Project manager: Shall be responsible to the Board of Directors. The project manager shall be responsible for guiding and co-ordinating the efforts of all parties engaged in implementing the project, obtaining necessary government approvals on contracts. He is to control the project organisation with the promoters as well as with other agencies and organisations interested in the project. The manager shall have some staff to assist him, especially in checking expenditures to date and determining the present and future cost overrun or under run so that the project manager can take or propose to the Board pertinent corrective measures.

The network shall cover the pre-construction phase of the project indicating major administrative processes, since experience shows that some of them have frequently involved lengthy delays. In other words, it shall include the aggregate activities to be carried out be the principal parties participating in the implementation process.

The project budget shall then be prepared. Part or periodic payments to contractors which might be made at the end of certain intervals (e.g. weekly or monthly) throughout the time horizon of project implementation, shall be made by summing up activity costs per unit of time, which may be a week & month, and computing the cumulative cost at the end of each time interval. For the activities that are in process and are contacted or sub-contracted, the assumption of a linear time cost activity relationship shall be used for the sake of simplicity. In other words, expenditures are uniformly distributed throughout the duration of the activity.

PROJECT SCHEDULE

After an investment decision is taken, the main machinery and long delivery items must be ordered out at the earliest, forming the first major step in implementation of the project. It is foreseen that an engineering consultant will be appointed for carrying out the detailed engineering including basic engineering and procurement assistance to the client. It is also assumed that reputed and experienced contractors with adequate resources viz., men, materials, tools and tackles etc. will be engaged for execution of the construction and erection work. The purchase packages for auxiliaries shall be kept minimum so as to reduce the co-ordination efforts to the minimum. A great deal of co-ordination is required for constructing/erecting the new units. This task is feasible, provided the major activities of the project are co-ordinated and completed in the duration specified to achieve the respective milestones in time.

STRATEGY FOR TIMELY EXECUTION

It is important to deploy a team of experienced personnel for project execution and select the external agencies with due care for rendering the services and supply of equipment for the project. The project activities must be identified, planned and scheduled, and the progress monitored for timely project implementation. All the inputs to the project including financial resources must be identified and their inflow planned and arranged in time.

However this tentative schedule, many activities

will begin in the 6th month after the start of the financing, using different locations and training some partial personal. This will turn possible the strength of the dynamics in the whole process as well as the shorting of the estimated timing.

In short, the following key factors would constitute the broad strategy for timely execution of all activities in a pre-determined manner as per schedule shown in the bar chart as to reach at a basis of regular production.

Early selection of an effective in-house technical team (TASK FORCE) by Government of Mozambique, headed by a Project Manager for planning and executing the project.

i)   Proper choice of external agencies such as consultants for Project Engineering., Machinery suppliers, Construction Agencies etc. keeping in view their reputation/past performance and working experience in their fields.

ii)   Adequate use of computer-based PERT/ CPM techniques for project planning, scheduling and monitoring.

3.6   Description of the initial project Flowchart.

The project implementation phase embraces the period from the decision to start the project to the beginning of the commercial production. It includes a number of stages including negotiations and contracting, project design, construction and start-up.

3.7   Description of the requirement of Land.

For this project the Government of Mozambique will participate with about 25.000 square meters of land. In selecting the land the following criteria should be followed.

1.   The land should be near to the main city of Maputo.
2.   The land should be properly connected by road.
3.   It should also have proximity to many end user clients.
4.   It should also meet various Government policy of
        achieving the social objectives.

2.08 Description of the requirement of Building.

The total requirement of building is as follows.
a. Adm. Building   700 Square Meters

b. Studios –3 (Three)   1.800 Square Meters
c. Other Misc. Building   300 Square Meters

Total Square Meters: approximately 2800 Square Meters.

Cost for construction of about US$750 per square meter (including air conditioning systems, primary electrical cabin etc.).

2.9   Description of some requirements for Plant & Machinery and their basis of selection.

The plant and machinery have been selected giving due consideration to the sophisticated nature of technology required. Detailed discussions were carried out with the foreign suppliers to ensure that the required capacities are practical with minimum capital and operating costs before the machinery were finally selected.

The machinery shall be selected from the most reputed foreign manufacturers of complete range of equipment. Keeping in view the following main factors:

a)   Past performance
b)   Existing machinery in Africa & abroad.
c)   Sales and service facilities in Africa

The main equipment is as follows.

STUDIO

1.   Post-production equipment – non-linear post-production, two ES7 stations.

2.   ENG – 3 DVCAM Camcorders

3.   Régie with 4 DXC-D30PK1 Digital Cameras - for cold light installation.

4.   Flexicart.

5.   Computer graphics suite.

6.   Sound room.

7.   MUST SYSTEM: Automatic Emission Managing System, with graphical information for no active emissions. (This system has been used by several French thematic channels)

Other technical data:

The whole installation is characterised by:

1.   Informational web with UTP level, 5 cables.

2.   Phone system.

3.   Specific furniture.

4.   Parking for about 200 places

Discrimination of Technical Data:

Studio

Cold lighting system.

4 Camera channels composed by:

2 DXC-D30PK Sony Digital Cameras

4 J18 Canon Zoom Lens

4	CA-537P Sony Camera Adapters
4	DXF-50 Sony Studio Viewfinder's
4	CCU-M3P Sony Control Camera Units
4	RM-7P Sony remote CCU controls
4	50M Sony Multicore Camera Sets
4	Canon Focus and Zoom control sets
4	Intercommunication Headphones
4	Special tripods

1 charriot plane and curve

1 Teleprompter with 2 reading systems (monitors for individual speakers) 1 Lighting control table with memory – ARRI
Different microphones with diverse directional characteristics Audio components
Video components
General Intercommunication system

Video Régie

1 Recorder/Player BETACAM equipment
1 video digital mixer with 12 tracks
1 DVE equipment of digital effects for 3 channels
1 TBC Frame Synchroniser
2 Oscilloscope and Vectorscopy for control of cameras
2 Synchroniser Generators with Changeover
1 Intercommunication system with 4 places
1 Audio and Video Matrix with remote controls Video distributors

Audio Régie

1 audio mixer with 24 tracks
2 audio monitors
1 cd player
1 cassette deck
1   DAT
1   Mini-Disc
1   Level Detector for Audio Stereo Stereo Audio Distributors
Audio compressors Audio effects Microphones

Video Post-Production
3 Sony ES-7 on DVCAM/AVID hybrid non linear edition stations 1 Conventional BETACAM Edition Suite 2:1
Multi-track audio (8) on hard disc, Sound Scape type

3 ENG SETS:
Sony DSR-300 PK DVCAM Camcorder Digital Compact Report projectors
Various support materials (Batteries etc.)

3.10   Description of the Water Requirements for the project.

The requirements of water, separately for various matters are given in the following table.

Circulating   : Nil
Make-up   : Nil
Process   : Nil
Drinking   : 3,000 LPD

3.11 Description of the Steam requirement for the project.

A. Steam requirements and steam balance   : N.A.

B. Capacity and type of boiler with detailed specifications   : N.A.

C. Steam and energy diagram : N.A.

D.   Total energy generated / purchased (converted into M. K. : N.A. Cal) theoretical requirement of energy (in M. K. Cal)
at the various consumption stations and expected actual requirement at these stations.

E. If alternate processed are available, comparative energy : N.A.
consumption figures for the various processes. If

the project is energy intensive, possibility of choosing alternate process in order to make the project less energy intensive.

F. Steps proposed to be taken by the company to improve : N.A.
energy losses efficiency and reduce energy losses (such as power factor improvement, power load management, optimising, illumination waste heat utilisation, etc.)

G. Scope for usage of solar / other renewable sources of   : N.A.
energy.

H. Any other measures contemplated in the direction of : N.A.
energy conservation and management.

3.123 Some information on Compressed air, fuel, etc.

Compressed Air

(a) Requirement   : N.A.
(b) Sources   : N.A.
(c) Arrangements proposed   : N.A.
(d) Cost at site with detailed calculations : N.A.

3.13   Details of the nature of atmospheric, soil

and water pollution likely to be created by the project and the measures proposed for control of pollution. Indicate whether necessary permissions for the disposal of effluent have been obtained.

There shall not be any atmospheric pollution likely to be created by the project as there is no machine which has combustion resulting into air pollution or any chemical process which may release any gases which may result into an air pollution.

However, the use of DG set shall cause a small amount of air pollution. Considering the size of DG set, the pollution is within the permissible limits.

MOZAMBIQUE TELEVISION NETWORK

THE INDUSTRY

CHAPTER - 4

THE INDUSTRY

4.01 Description of the Television Industry in general.

The television industry can be described in the following broad categories. 1. Introduction

2.   Cable Networks

3.   DTH

4.   TV-media

5.   Earnings drivers

6.   Outlook

The detailed discussion in each of the categories is given below.

Introduction

The growing popularity of TV as a communication medium has resulted in the TV media sector undergoing a rapid transformation. From the black and white days of state controlled TV Station, to the highly colourful tunes of Channel V and MTV, the medium has certainly undergone a phenomenal change. Given its popularity, percentage ad spend has also increased proportionately on this medium.

Media pie (%)

	1995	1997
TV	62.5	68.8
Radio	20.9	15
Press	16.6	16.2

Source NRS

Entry of new channels post 1991

All over the world the telecasting has witnessed entry of new channels to cater to the various needs of
world audiences. Channels have been launched in
English as well as other regional languages. In many
countries of the world till 1991, the state owned TV
Station ruled the roost, as other players were not
allowed to uplink and broadcast. However channels
such as CNN, Star TV and BBC, which were offshore
companies, could circumvent these regulations
and telecast their programs into any country of the

world. Cable operators then relayed the same and made it available to the common man through the cable television network.

Like many other countries, the State machinery controlled television. It was used as a propaganda tool for the party in power, with the opposition always at the receiving end. The customer had very little choice. The first steps towards more user choice began during the 1980s, which had to be telecast to a wider audience. TV Station used satellite channels for the telecast and the T.V. network was launched as an international channel.

The sports telecast by Channel 9 in 1985 and the Gulf War in the late eighties all played small but important cameos in educating the international viewer. With liberalisation in 1992 and crumbling tariff barriers televisions (read as colour TVs) became more easily available. The media revolution had started.

Major satellite channels avidly watched by viewers are Star TV, Sony TV, Home TV, BBC & CNN. There are other regional language channels which are major players in their respective territories.

Most of the channels that could not attain popularity rapidly suffered, since their advertisement earnings were not sustainable. The first round of the media wars is over. Management changes, i.e.

original promoters selling out to new management with deeper pockets, has become the order of the day. Alliances like the famous ESPN Star Sports arrangement also made headlines. Given the global trends of mergers and acquisitions, further consolidation is likely. Alliances and mergers make sense when the partners complement each other, like BBC and Discovery launched Animal Planet, CNBC and ABNI came together to launch a business channel called CNBC Asia.

Cable Networks

Antennas set up by either the end user or the cable operator receives the signals transmitted by the satellite. Local cable operators lay their own cables, set up control rooms, which can telecast 40 or more channels over a limited area. They charge the household a one- time connection charge of about US$.10 per point and a recurring monthly charge ranging between US$.1 to US$.5.

Initially, this was done in a very unorganised manner. The business required local knowledge and contacts, so every locality had its own cable operator. Collection was critical for the cable operator. For the end user, quality of telecast and a complete lack of standards became an issue. This lead to a shakeout and the formation of cable companies with money power which in turn tied-up with the local and small cable operators. Cable companies charge about US$1 per month to the local cable operators and support them with training and other infrastructure inputs. The business is immensely capital intensive and takes a long time to break even.

In many countries the operations of cable operators are regulated under the Cable TV Act which ensures that pornographic materials and other materials which are against culture and values or are detrimental to national interests do not get telecast. Recently this act has been amended to include foreign channels also.

Direct To Home

DTH is a new technology that circumvents the cable operators by directly delivering a bundle of channels to the end user. DTH involves transmission of encoded audio/ video signals (Ku band) via satellite. The end user needs an antenna to receive the signals and a decoder (set top box) to unscramble the encrypted signals. DTH services elsewhere in the world are Echostar and DirecTV (USA) and BskyB (Europe). Rupert Murdoch of Star TV fame owns BskyB.

The size of the antenna in DTH will be 1.5-2 ft in diameter, making it easy to install and transport. In conventional cable, since signals are in C band, an

8ft- diameter antenna is needed. The basic difference in the business model is the hardware costs in DTH. In a cable system, the user pays a one time connect fee and monthly rentals, while in DTH he has to invest in hardware.

The antenna will cost about US$200-300 and decoder will cost about US$200. The African viewer might be reluctant to incur such heavy installation costs. Quality of telecast in DTH is superior to Cable TV and viewer can receive up to 200 channels.

DTH will result in restructuring of the cable television industry. It will become imperative to have cash reserves to withstand the technology threat. Up gradation to fibre optic backbone will become necessary. A fibre optic network will cost about US$0.5mn per km as compared to US$0.1mn per km for coaxial cable. The stage is now ripe for consolidation.

TV Media features

In the Broadcasting business, it is only the industry leader who makes sizeable profits. The business is a game of asymmetrical payoffs. For instance, the top 5 channels account for 90% of ad spend.

Urbanisation and TV penetration is related. This may be due to the popularity of cable television that has resulted in increased colour TV sales. Rural penetration is low, although growing at a fast pace, because of dearth of specific program content to cater to that segment.

Liberalisation has resulted in the world viewer becoming more aware and conscious. This has resulted in the customer having more choice with the entry of a number of companies in different segments. Competition has resulted in companies increasing their marketing spend significantly.

Popularity of TV media is becoming higher. Increasing TV penetration leads to a reallocation of advertisement budgets with higher allocation for television at the cost of other medium.

TV channel operators use different business models to generate revenues. The critical component of any channel is the quality and type of programs they telecast. This determines their popularity, which in turn determines amount of advertisement revenues they can generate. They can do any one of the following:

Buy programming rights of program software from outside and collect advertisement revenue on their own. This model is followed by several TV companies, wherein they have a separate company in their fold, which develops all the content. The advantage is that re runs of serials/ programs become very profitable.

Selling time space to the producers for a fixed charge. Producers in turn are free to book advertisements at their own rates (there is an understanding on the time allocated for advertisement) and collect revenue. This is the basic model for many TV companies in which they sell prime time slots. The rights continue to be vested with the producer.

Earnings drives

The key factors that drive sector revenues are

Television penetration: Since the medium is television, increased television penetration will imply higher viewer ship. This will translate into higher advertisement spend allocation. This will also imply higher software production and demand for new programs.

Competition from other satellite channels would have an adverse impact on advertisement revenues, as advertisers have more choice in allocating ad budgets.

Government policies can have a big impact on the fortunes of the entire industry. When the DTH bill is passed in any country then, it will trigger a restructuring of the cable business.

Launching new channels targeted at specific seg-ments, like regional channels within any country other areas having large pockets of ethnic popula-tion would lead to revenue growth. This will entail significant initial outlays.

Depreciation of the local currency would increase revenues as most of the program/ software compa-nies export the programs overseas and payments are dollar denominated.

Advertisement revenue

As mentioned earlier, this is the primary source of income for TV channel operators. This revenue is directly co-related with the reach and viewer ship of a channel. Any channel's popularity depends on good quality programs, which is the software con-tent. The business requires enormous initial invest-ment in programs and revenues follow only with a time lag after the channel receives a minimum viewer acceptance.

Outlook

The sector has latent potential for growth on back of the exponential growth of cable TVs during the last 5 years. Television penetration in Africa is extremely low as compared to other developing countries like Malaysia, Pakistan, etc. in Asia. The number of channels has increased, implying higher demand for software programs.

Advertisement revenues, which are the barometer of channel popularity, will get dispersed over several competing channels. A shakeout is likely in both the channel and cable TV sectors. The biggest beneficiaries will be the content providers or the software houses. They will control the intellectual rights to the key element driving any channel's popularity.

Direct-to-Home, Digital Terrestrial Transmission and Conditional Access Cable Delivery have emerged as new delivery mechanisms. Breakthrough in technology would help open up avenues for these channels.

MOZAMBIQUE TELEVISION NETWORK

MARKETING

CHAPTER - 7

MARKETING

7.1   Description of the commercial viability of the project with regards to revenue generations.

Estimations shows that, after about 5 years of regular working, the CPLP Cultural Television Channel will be able to start exporting a great quantity of services and videos to Africa, Europe and South America. (See financial projections)

The economic feasibility of the project will occur with the commercialisation of several products, like:

1.   The sale and production of cultural programmes to various institutions like:
-    BBC
-    TV Culture Brazil
-    RTP
-    Discovery
Etc.
All over the world commercial networks for tel-

evision cultural programmes have been created. Several television channels in different countries have regularly acquired programmes by producers spread out all over the world. Not only, several international organisations, like the World Bank and the United Nations, including FAO and Unesco, among many others, need programmes for public education, like programmes oriented to alert and to educate people concerning endemic and epidemic diseases. AIDS, malaria, tuberculosis are a very few examples.

These world institutions have made a great effort to develop television programmes with humanitarian objectives. But, in Africa – undoubtedly the continent most needed of such programmes – there is no television station, in present times, with capability to make face to such a need.

Therefore, the television teams that had been responsible for such programmes are, practically in its totality, placed in countries of the called First World, strange to the local population's concrete reality.

Programmes focusing new agricultural techniques or even oriented to agricultural, commercial and industrial education are essential elements of such a repressed demand.

The price for each institutional campaign, with an

average of 10 films produced per campaign, is of about USD.15.000$ and the capacity of the CPLP Cultural Channel will be as follows:

First Group* - from 2 to 4 campaigns per month.
Second Group – from 1 campaign in two months to 1 campaign per
month

This represents:
First Group   a potential annual income of about US$.540.000
Second Group a potential annual income of about US$.90.000

*First Group: Angola, Mozambique, and Mozambique. Second Group: Guinea-Bissau, Sao Tome e Principe, East Timor (Lorosae). Reference Group: Brazil, Portugal.

Obs. In the Second Group, for the first six years of operation it is estimated a difference of potential annual income between the Angola unit and Mozambique's and Mozambique's respective units, from US$.540.000 to about US$.378.000.

2. The sale and production of no cultural programmes to other television channels, like:
-        series (novels)
-        talk-shows Etc.

Many countries of the region need to produce television programmes, but do not have capability to do it. Thus, they seem themselves obliged to search expensive productions in Europe. With only one programme sold-three daily hours-the channels of the First Group* would receive in incomes the equivalent of about two times of the whole investment. The Second Group countries will not have conditions at the beginning to product cultural programmes to other television channels.

*First Group: Angola, Mozambique, and Cape Verde. Second Grope: Guinea-Bissau, Sao Tome e Principe, East Timor (Lorosae). Reference Group: Brazil, Portugal.

The price for this type of television programme is:

Auditory programs   US$.12.500 per programme
Talk-shows   US$.4.000 per programme
Interviews   US$.4.000 per programme

The capacity of production, of these programmes by the countries of the First Group will be:

In the FIRST PHASE:
Talk-shows   2 per month

Interviews   4 per month
In the FINAL PHASE:
Auditory programmes   6 per month
Talk-shows   20 per month
Interviews   20 per month

Thus, the potential income will be:

FIRST PHASE
Talk-shows   US$.96.000 per year
Interviews   US$.192.000 per year

FINAL PHASE
Auditory programmes   US$.900.000 per year*
Talk-shows   US$.960.000 per year*
Interviews   US$.960.000 per year*

* the same difference as showed above.

The price of the novels is much higher. For each complete novel, with about 120 chapters, the price is about US$.2.500.000 and the capacity of the First Group of the CPLP Cultural Channel (in the final phase) will be of one novel per year.

3. There is the possibility of commercial use of the transmission time beyond the six hours reserved to culture and education.

The conventional day in television is of 18 hours,

of which only 6 hours would be no commercial. We would have, therefore, 12 hours of no cultural transmission, which should be freely commercialised.

Even the period of six hours- divided into two sections of three hours, the first one dedicated to culture and second section of three hours to education-has a great potential for sponsoring.

Each commercial hour of transmission can include up to 12 minutes of advertising. The price for each 30 seconds of advertising is:

Noble time   from 6PM to 11PM   a b o u t US$.400 each 30"
Normal time   rest   about US$.200 each 30"

Concentrating the commercial programming in the noble time. With 4 hours of transmission in this period, we would have:

Educational television: from 7AM to 9AM
From 3PM to 4PM
Cultural television: from 4PM to 7PM
Commercial television: from 9AM to 3PM
From 7PM to 1AM

Thus, the commercial period would comprehend 6 hours in normal time, 4 hours in noble time and 2 hours in normal time again.

Begin 12 minutes per hour for advertising, the commercial period could have:

Noble time   48 minutes   24   films   US$.9.600 per day

Normal time   96 minutes   48   films   US$.9.600 per day

Total of the potential income............ US$.19.200 per day

> or US$.576.000 per month
> or US$.6.912.000 per year

It is believed that in the first phase of the project (after one year), the CPLP Cultural Television Channels of the First Group will be able to start with an income from commercial advertisement of about US$.1.000.000 per year-value which is predicted to increase in the follow months.

4. The rent of the studios to television teams of other countries can be another source of incomes. Many producers who cover events in Africa need to move many times to their countries of origin during the video works, because there is not technical support in Africa. The same phenomenon happens with the cinematography and the journalism productions. It is not difficult to imagine, for example, the serious problems journalism teams have had, for example, with essential components like batteries, lighting, electronic components etc., which only can be easily find in Europe.

The period of renting of a television studio is of 12 hours. Each period has price of about US.4.000. After the first year, the studio should be rented for 12 hours per each period of 3 days. So, the rent per month, in this period (First Phase), will be able to generate incomes of about US$.120.000 per month in the First Group.

5. Support services to other television networks in all areas, including novels, mini series, docdramas series etc.

Many countries, principally in Africa, do not have technical conditions to develop this kind of programmes, but they have a strong internal repressed demand in this sense.

Each team, abroad, has a price of US$.1.000 per day (two people), after the costs of dislocation, hotel and meals. In the first months (first phase) it will be organised one team for works in other countries. In the final phase it is predicted to have up to 5 teams with such function. The capacity of each team is of about 20 days per month. Therefore, the support to other television networks-when took in its full potentiality-will be able to represent up to 20 days per month in the first phase. This represents a potential income of about US$.240.000 per year, and 100 days per month of works in the final phase, signifying about US$.1.200.000 per year of incomes, always referring to the First Group countries.

6. Colloquies, seminars, videoconferences and meetings of different natures.

The building of the CPLP Cultural Television Network will have a medium size auditorium, (First Group), with all conditions to receive seminars, colloquies and meetings of the most diverse nature. Such seminars, videoconferences etc. are important not only for the increase of the incomes, but also to attract specialists of the most different areas, being an important element for the development of all region as well as for the diffusion of the local, regional and continental cultural values.

The meetings, seminars and colloquies attract, in average, about 300 people per event. The price-excepting meals, hotels and transportation-per each participant is of about US$.50 per day. It will be a capacity for up to four events of this type per month, what could represent an income of about US$.60.000 per month or US$.720.000 per year (final phase-after 5 to 7 years).

The project should also turn possible.
-        classes of professional formation in diverse disciplines
-        support for a multimedia high technology centre (First Group)
-        support for a multimedia high cultural centre (Final Group)

MOZAMBIQUE TELEVISION NETWORK

PROFITABILITY & CASH FLOW

CHAPTER - 8

PROFITABILITY AND CASH FLOW

8.1   Estimation of cost of production and working results for the first five years of operation.

The estimated cost of production of working results for the first five years of operation are given in the chapter "Financial Projection".

8.2   Cash flow statement for the company as a whole, for five operating years of the project based on the estimates of working results.

A detailed cash flow statement for the company as a whole for five operating years is given in the chapter of "Financial Projection".

8.3   Projected balance sheet for five operating years for the company as a whole.

The balance sheet for five operation years for the company as a whole is given in the chapter of "Financial Projection".

MOZAMBIQUE TELEVISION NETWORK

ASSUMPTIONS

CHAPTER - 9

ASSUMPTIONS

1.   The Mozambique Television Network shall generate the income from the following sources.

a)   Sale of cultural programmes.
b)   Sale of commercial programmes.
c)   Sale of advertisement time during transmission of TV programmes

d)   Hiring of studios.
e)   Supply of technical services
f)   Hiring of conference hall

The detailed assumptions for each of the above mentioned activities are as follows:

a)   Sale of cultural programs:

I.   It is assumed that four cultural programs shall be produced per month in the first year, which can be sold to other countries. This
figure will increase to 5 programs per month in the second year, six programs per month in the third year and so on.
II.   It is assumed that the selling price shall be US$.105 per program.

b)   Sale of commercial programs:

I.   The commercial programs consist of talk shows, interviews and auditory programs.
II.   It is assumed that two talk shows shall be produced per month in the first year, three talk shows shall be produced in the third year, four talk shows per month shall be produced in the third year and
so on.
III.   It is assumed that the selling price of one talk show program shall
be US$.2800.

IV.   It is assumed that four interviews shall be produced per month in the first year, five interviews shall be produced in the third year, six interviews per month shall be produced in the third year and so on.

V.   It is assumed that the selling price of one interviews program shall be US$.2800.

VI.   It is assumed that no auditory programs shall be produced in the first and the second year. Only when the people have two years of experience then in the third year one auditory program shall be produced per month. In the fourth year two programs shall be produced per month and so on.

VII.   It is assumed that the selling price of one auditory program shall be UD$.8750.

c) Sale of advertisement time during the transmission of T.V. programs:

I.   There shall be a total transmission of 8 hours per day in the first year. It will go on increasing by two hours in the second and subsequent years.

II.   Out of which advertisement shall be available for programs with transmission period of four hours in the first year. The programs in
which advertisements could be available shall increase by two hours per year in the second and subsequent years.

III.   In one hour T.V. transmission, 12 minutes of advertisement shall be allowed.

IV.   Out of total T.V. time 33 % time shall be considered as prime time and remaining 67% shall be considered as non prime time, in the first, second and third years. From the fourth year onwards the prime-time advertisements shall remain constant.

V.   The advertisement rates shall be US$ 560 for one minute of prime time advertisement and US$ 245 for one minute of non prime time advertisement.

d)   Hiring of studios:

I.   It is assumed that the studios shall be taken on hire for a shift of 12 hours, two times in a month in the first year. It will increase to three times in a month in the second year.

II.   The rent per day shall be US$2800.

e)   Supply of technical services:-

I.   It is assumed that one team, consisting of two technically qualified people shall be available in the first year. In the second year two such teams shall be available. In the third year three such teams shall be available and so on.

II.   It is assumed that in the first year the team shall be hired for 10 days in a month. In the second year the teams shall be hired for 12 days in a month. In the third year the teams shall be hired for 14 days in a month and so on.

III.   The rate of hire shall be US$.700 per team per day.

f)   Hiring of conference hall :-

I.   It is assumed that the conference hall shall be taken on hire for two times in a month in the first year, three times in a month in the
second year and four times in a month in the third year and so on.
II.   The hire charges shall be US$.350 per conference.

2.   The unit is planning transmission of three hours of educational programs, three hours of cultural programs and two hours of commercial programs per day in the first year. Thus, there shall be a total transmission of eight hours of programs per day in the first year. The total transmission hours shall increase by two hours per day in the second and subsequent years. Thus, in the second year there shall be total transmission of 10 hours per day and in the third year there shall be a total transmission of 12 hours per day. The cost of software production is assumed at US$.1750 per hour.

3.   The cost of miscellaneous consumable items is assumed at about US$.25200 in the first year. This will increase by 10% in the second and subsequent years.

4.   The cost of stores and spares consumed is assumed at US$.29400 in the first year. It will

increase by 10% in the second and subsequent years.

5.   The cost of repairs and maintenance is assumed at 0.1% of the total cost of plant and machinery in the first year. This cost will increase up to 0,3% in the fifth year.

6.   It is assumed that the unit shall use 1.000 units of power every day or 365.000 units of power in the first year. The consumption of power shall increase by 10% in the second and subsequent years. The cost of power is assumed at 3.5 cents for one unit.

7.   It is assumed that the unit shall consume 700 litters of diesel per month. The cost is assumed at 33 cents for one litter. It will increase by 10% in second and subsequent years.

8.   The unit shall have a technical staff of 35 people and administrative staff of 15 people.

9.   Depreciation is calculated on reducing balance method. The rate of depreciation is taken at 10% for building, 25% for plant and machinery, 10% for furniture and fixtures and 20% on vehicles.

10.   Advertisement expenditure is assumed at 0.2% of the sales.

11.   It is assumed that the stock of raw material and stores shall be 30 days of consumption, receivables shall be 45 days sales and creditors shall be 15 days purchases.

12.   The term loan shall be for a total period of 25 years. It shall have a moratorium of five years. It shall be repayable in 20 equal yearly instalments.

PROJECT REPORT

OF

SAO TOME CULTURAL TELEVISION NET-WORK

AT

SAO TOME E PRINCIPE

AFRICA

PROMOTERS

REPUBLIC OF SAO TOME E PRINCIPE

AFRICA

SAO TOME TELEVISION NETWORK

INTRODUCTION

CHAPTER - 1

INTRODUCTION

1.1   Description of the organisation through which this project is being taken up.

The CPLP is an international organisation, created in 17 of July of 1996, with Headquarters in Lisbon and consisting of the following State Members: Angola, Brazil, Cape-Verde, Guinea-Bissau, Mozambique, Portugal, Sao Tome e Principe and East Timor (Timor Lorosae).

The CPLP has as its objective the politic-diplomatic relationships between its Members, mainly referring to the United Nations and the World Bank, the relationship EU/MERCOSUL and the accomplishment of the Europe-Africa Summit.

The CPLP is also dedicated to the co-operation, particularly in the economic, social, cultural, legal and technical-scientific matters; as well as to the projects of promotion and broadcasting of the Por-

tuguese Language, nominated to the improvement of the International Institute of the Portuguese Language and the creation of a Bibliographical Fund.

All its decisions are taken by consensus.

2.2   Description of the main reasons for setting up an African Cultural Television Network.

Africa is a continent submerged into dramatic problems. All these humanitarian problems have in the education and culture their most direct and objective root.

The African countries speaking Portuguese are among the African countries with the lowest level of development.

The whole situation is so dramatic that the implementation of the CPLP Cultural Television Network is more than urgent.

It is hoped that with this Television Network the people shall have better understanding of various diseases and other problems of life. e.g. AIDS is prevalent disease in Africa. But as the people are illiterate the information about the disease cannot be spread by any other method. In such cases the Television Network becomes a very effective audio-visual means of communication.

The project also intends to be a base of a massive production of health campaigns, specially to reinforce the combat against the AIDS, the infect-contagious diseases- endemic or epidemic-through a process of global education of the local and regional populations.

The CPLP Cultural Television Network is a project for peace and its objective is not to compete with local television channels already installed, but to interchange with them, promoting a future global web dedicated to knowledge and development.

1.03 Description of the main objects of the project.

The major objective of the project is to assist the CPLP countries to develop the first African television network channel oriented to cultural and educational themes with the follow orientation:

-       diffusion of the Portuguese language
-       education in all areas
-       reinforcement of the local and regional cultures
-       formation of technical personal on communication and tele-education
-       formation and information for a better participation of the CPLP nations in the process of global development
-       to support the implantation of the Institute

of Portuguese Language

The development of such a project will create an irradiation pole from the countries with the project implementation to the whole African continent, constituting an African network of research and culture, discovering and valorising different cultural aspects and, consequently, reinforcing identity factors.

With the stronger cultural identity, it is expected to have lower levels of violence in social behaviour as well as a relevant increase of the educational performance in all its sectors.

Systematic educational objectives are complementary to the general orientation of this project, which is cultural in its fundament.
A very large positive impact is also to be expected with respect to several sectors of the economy, greatly improving the overall quality of services.

1.4   Description of the mission status of the project.

The mission of the project is:

-       to rescue lost values and cultural information
-       to reinforce historical elements

- to rescue old artworks
- to give a new value to new artworks
- to communicate practical and objective information on health
- to communicate practical and objective information on agriculture
- to communicate practical and objective information on economy

1.05 Description of the philosophy of the project.

The CPLP Cultural Television Network will have in its philosophy, as a support to its mission, the follow fundaments:

- to work with local people
- be focused on quality
- intercommunication

Three words can define the CPLP Cultural Television Network philosophy:
People, quality and intercommunication

Following these three key words of its philosophy, a single sentence also defines its mission: "A Continent of Culture".

Two main sectors characterise its global structure:
1. Cultural Sector
1.1.   To develop programmes on:
- Music

- Architecture (also vernacular)
- Archaeology
- Literature and poetry
- History
- Tourism
- Dance
- Plastic arts
- Astronomy
- Science in general
- gastronomy
- health
- anthropology
- local cultures
- others

1.2   The programmes can be of all types, the most important is to be attentive to communication. All programmes must reach the highest possible number of spectators.

2. Educational sector.
2.1   Systematic education in short practical courses.
2.2   Educational series with practical and objective information on:
- health care
- family
- agriculture
- managing
- services
- civil construction

As the CPLP Cultural Television Network must not be supported by the respective governments in the future, it will have a daily time of free commercial programmes up to 12 hours. This is a good solution to pay its costs and support its development.

1.06 Description the magnitude of the project.

The project proposes a Television transmission time of eight hours per day in the first year. It will increase by two hours per day in the second and subsequent years.

The project also envisages training of local people in all the related areas of Television. It is proposed to develop a local pool of talent which can be utilised not only in this particular country but also in other countries of Africa.

The total cost of the project of the Sao Tome Cultural Television Network is US$.13.200.000. Out of this it is proposed to obtain a soft loan of US$.11.000.000. The remaining amount of US$.2.200.000 shall be contributed by the Government of Sao Tome in terrain for the installation of the respective building, infrastructures and logistic support.

For the purpose of the Sao Tome Cultural Television Network a company shall be formed with the participation of the Government. The Contribution of the Government in the form of share capital shall be US$.4.500.000.

The term loan shall be on soft terms. The total period of the loan shall be 23 years. Out of which five years shall be a period of moratorium. The entire loan shall be repaid in the remaining 17 years, in 17 equal yearly instalments.

The Government of Sao Tome will give the sovereign guarantee for the referred loan.

SAO TOME TELEVISION NETWORK

THE COUNTRY

CHAPTER - 2

THE COUNTRY

2.01. Description the level of development of Sao Tome.

This small poor island economy has become increasingly dependent on cocoa since independence over 20 years ago. However, cocoa production has substantially declined because of drought and mismanagement.

The resulting shortage of cocoa for export has created a persistent balance-of-payments problem. Sao Tome has to import all fuels, most manufactured goods, consumer goods, and a significant amount of food. Over the years, it has been unable to service its external debt and has had to depend on aid and debt rescheduling. Considerable potential exists for development of a tourist industry, and the government has taken steps to expand facilities in recent years. The government also has attempted to reduce price controls and subsidies, but economic growth has remained sluggish. Sao Tome is also optimistic that significant petroleum discoveries are forthcoming in its territorial waters in the oil-rich waters of the Gulf of Guinea.

OFFICIAL NAME: Democratic Republic of Sao Tome and Principe
CAPITAL: Sao Tome
SYSTEM OF GOVERNMENT: Multiparty Republic
AREA: 963 Sq Km (372 Sq Mi)
ESTIMATED 2000 POPULATION: 148,700

LOCATION & GEOGRAPHY: Sao Tome and Principe are islands located in the Gulf of Guinea off the coast of West Africa. Both islands are of volcanic origin and are characterised by many craters and lava flows. Sao Tome has ten peaks over 1,067 metres (3,500 feet) while Principe has a larger and flatter area than Sao Tome. Streams radiate throughout the forest clad mountains to the sea on both islands. Major Cities (pop. est.); Sao Tome 43,400 (1991). Land Use; pastures 1%, agricultural-cultivated 39%, forested and other 60% (1993).

CLIMATE: Sao Tome and Principe has a tropical climate characterised by hot and humid conditions that are influenced and modified by the cold Benguela current as well as by altitude. The dry season is from June to September and the wet season from October to May. Average annual precipitation varies from 500 mm (197 inches) on the south-west mountain slopes to 1,000 mm (394 inches) on the northern lowlands. Average temperature ranges in Sao Tome are from 21 degrees Celsius (70 degrees Fahrenheit) to 31 degrees Celsius (88 degrees Fahrenheit) with the temperature of the interior's higher altitude around 20 degrees Celsius (68 degrees Fahrenheit) .

PEOPLE: The principal ethnic majority is formed by Africans. Other ethnic minorities include Angolares who are descendants of Angolan slaves and the Portuguese.

DEMOGRAPHIC/VITAL STATISTICS: Density; 123 persons per sq km (319 persons per sq mi) (1991). Urban-Rural; 40.5% urban, 59.5% rural (1988). Sex Distribution; 49.4% male, 50.6% female (1991). Life Expectancy at Birth; 64.0 years male, 67.0 years female (1990). Age Breakdown; 38% under 15, 22% 15 to 29, 17% 30 to 44, 12% 45 to 59, 8% 60 and over, 3% unspecified (1989). Birth Rate; 38.0 per 1,000 (1992). Death Rate; 8.0 per 1,000 (1992). Increase Rate; 30.0 per 1,000 (1992). Infant Mortality Rate; 58.0 per 1,000 live births (1992).

RELIGIONS: Mostly Christians with 84% of the population Roman Catholic while the remainder follow local native tribal beliefs.

LANGUAGES: The official language is Portuguese

which is spoken in a heavy Creole dialect.

EDUCATION: Aged 25 or over and having attained: no formal schooling 56.6%, incomplete primary 18.0%, primary 19.2%, incomplete secondary 4.6%, secondary 1.3%, higher 0.3% (1981). Literacy; literate population aged 15 or over 28,114 or 54.2% (1981).

CURRENCY: The official currency is the Dobra (Db) divided into 100 Centimos.

ECONOMY: Gross National Product; USD $41,000,000 (1993). Public Debt; USD $225,800,000 (1993). Imports; USD $30,400,000 (1994). Exports; USD $6,500,000 (1994). Tourism Receipts; USD $1,000,000 (1990). Balance of Trade; USD -$23,900,000 (1994). Economically Active Population; 49,216 or 41.0% of total population (1991). Unemployed; 22.0% (1994).

MAIN TRADING PARTNERS: Its main trading partners are Portugal, Angola, the Netherlands, the USA, Germany and the UK.

MAIN PRIMARY PRODUCTS: Bananas, Cocoa, Coconuts, Coffee, Fish, Palm

Kernels, Timber.

MAJOR INDUSTRIES: Agriculture, Food Processing, Light Construction,

Timber Products.

MAIN EXPORTS: Cocoa, Coffee, Copra, Palm Kernels and Nuts.

TRANSPORT: Railroads; nil. Roads; length 380 km (236 mi) (1988). Vehicles; cars 1,774 (1975), trucks and buses 265 (1975). Merchant Marine; vessels 3 (1990), dead weight tonnage 1,172 (1990). Air Transport; passenger-km 6,100,000 (3,790,000 passenger-mi) (1985), cargo ton-km 100,000 (68,490 short ton-mi) (1985).

COMMUNICATIONS: Weekly Newspapers; total of 2 with a total circulation of N/A . Radio; receivers 31,000 (1994). Television; 21,000 (1994). Telephones; units 2,400 (1993).

MILITARY: N/A. Military expenditure accounts for 1.6% (1980) of the Gross National Product (GNP).

SAO TOME TELEVISION NETWORK

THE PROJECT

CHAPTER - 3

THE PROJECT

3.1   Description of the initial steps to be taken for the implementation of the project.

The first objective of the CPLP Cultural Television Network is to install its structures in the following countries:

- Sao Tome e Principe
- Cape Verde
- Mozambique
- Guinea-Bissau
- Angola
- Timor

Thus, these six countries will structure the basic network of the whole project.

Angola, Mozambique and Cape Verde will form a FIRST GROUP; Sao Tome e Principe, Guinea-Bissau and East Timor will form a SECOND GROUP; and, finally, Brazil and Portugal will form a called REFER-ENCE GROUP.

However, the project will start at the same time in all countries. A more precise description of the meaning of each group is made below.

The basic physical components of the project are:

1.   The creation of companies, in each of the States of the CPLP, oriented to the objectives above referred.
2.   The State of each country – who will be the responsible and guarantee of the payment of the financing – will also have a participation in the companies.
3.   All strategies will be oriented to lowest costs in long term.
4.   The whole project will be developed by phases.
5.   The timetable of transmission will be defined case by case.
6.   The project will include:
6.1.   a complete plan for an equipment network.
6.2.   a complete plan for a software network.
6.3.   all procurements
6.4.   equipment transportation and installation
6.5.   all configurations
6.6.   creation of the infographic system
6.7.   technical formation of the personal

6.8.   architecture and engineer projects for the building
6.9.   construction of the building organised in phases

1.2   Description of the procedure by which the necessary programmes will be produced for transmission on the channel.

The television programs are known as software in this field. The unit is planning transmission of three hours of educational programs, three hours of cultural programs and two hours of commercial programs per day in the first year. Thus, there shall be a total transmission of eight hours of programs per day in the first year. The total transmission hours shall increase by two hours per day in the second and subsequent years. Thus, in the second year there shall be total transmission of 10 hours per day and in the third year there shall be a total transmission of 12 hours per day. The cost of software production is assumed at US\$.2.500 per hour. The amount of US\$.2.500 per hour is taken based on cost of production of such programmes in Portugal.

To contract a similar system from other television channel, only for one country, for a transmission period of 6 hours daily, the cost in, market (eg. Portugal*) is of about US\$.500.000 per month. This cost represents a total of about US\$.6.000.000 per year. In 16 years, excluding the first year of implementation, the costs for contract the six daily hours of transmission from other television channel, in present market price, is of about US$.100.000.000. The six countries would have an estimated cost of about US$.600.000.000 in a 16 years period of time.

These costs are based on the prices presently used in Portugal, as for example the programme Hora Viva – Seguranca em Directo, transmitted every day, excluding weekends, from 7 to 10 AM.

Another example is the channel Canal Noticias Lisboa CNL. Only in its first year of installation, it was expended about US$.10.000.000, and the equipment as well as the building were not property of the channel, but rented.

In fact, however the project will be implemented by phases, after 5 to 10 years of regular work, it is predicted to have reached a similar value in incomes as the examples above.

*value took in the Portuguese market, below the average of the European prices in about 30%.

The first ten months of works will be oriented to:
-        design of the equipment and software struc-

tures
-       start the first productions
-       elaboration of the projects of architecture and engineering
-       procurement and acquisition of all equipment
-       construction and supervision of the works
-       installation and configuration of the equipment

So, the first year must be used to the implantation of the project. However, as to accelerate the chronogram, some operations should be started before this timing, temporarily installed in a different building.

The first phase will be oriented to:
-       video productions (external and internal)
-       edition
-       elaboration of cost programmes

3.3   Description of the requirements of Power, Fuel and Water. POWER
It is assumed that the unit shall use one thousand units of power of everyday. Thus, the unit shall use 365.000 units of power in a year. The costs of power is assumed at five cents for one unit. The consumption of power shall increase by 10% every year.

FUEL

The unit shall use D.G. set for generation of electricity, when there is an interruption in the supply. The hours of disruption in the normal supply can not be predicted accurately. Hence, the hours for which the D.G. sets shall be used also can not be predicted accurately. However, it is assumed that the unit shall utilise one thousand litters of diesel per month. The cost is assumed that 33 cents for one litter. Hence, the total cost for one thousand litters shall be US$.330. The consumption of diesel shall increase by 10% every year.

WATER

The unit does not need water for fixed commercial operation. However, it shall have a strength of 50 people as employees. Water shall be required for drinking and sanitation purposes. It is assumed that 3000 litters of water shall be required everyday. The cost of obtaining this water is assumed at US$.10 everyday. The consumption of water shall increase by 10% every year.

3.4   Description of the manpower requirement of the project and description of the steps taken by the company to train the manpower.

The total staff strength shall be 17 people in the first year.

The system of modern TV Station is designed to achieve sustained operating efficiency and transmission. This, of course, entails a certain degree of sophistication in the production and transmission. However, considering the socio-economic situation in Sao Tome, a reasonable balance has to be struck to obtain optimum performance and at the same time create gainful employment. While working out the manpower requirement for this project to be kept on direct rolls of the company totalling 50 staff and operators, the above consideration have been kept in mind.

Manpower requirement

The direct manpower required for the proposed Unit is about US$.1.800.000 per year. The manpower requirement as indicated in this chapter has been planned keeping in view the following guidelines:

Effective co-ordination among the various departments. Judicious distribution of responsibilities. Capacity utilisation of the TV Station with optimum manpower.

Details of manpower requirement for the TV Station is given in the financial section.

In line with the prevailing practice, the Security guards, office peons and unskilled labour etc., are normally employed on contractual/casual basis, and their cost has to be included in the Factory over-heads. However in this report provision has been made to employ the above staff and their salaried have been included.

The various departments proposed shall be under the direct responsibility of the Station Director. He shall be assisted by Deputy Directors, who will look after the complete TV Station and its various day to day activities and Marketing. They shall be responsible for achieving the envisaged targets and sales forecasts. They shall be assisted by a team of Managers from production, marketing and accounts.

To run this project a labour force which shall usually be composed of unskilled and skilled workers shall be employed. The first are those who do not undergo any king of specific training or education, while the latter have to do so in order to master their jobs.

When evaluating an investment project from an employment point of view, its impact on both unskilled and skilled labour has been taken into account. Not only direct employment, but also indirect employment has been considered. Direct employment refers to the new employment opportunities created within the project; indirect employment concerns job opportunities created in other projects linked with the project which is being formulated.

The implementation of large and sophisticated projects generally contributes to the development of local skills and capabilities in a country. Furthermore, they help to change traditional values, attitudes and the behaviour of the society, to build up an enterprising spirit among the people, to develop a desire for changing and improving the existing conditions of life, to introduce better work discipline and thus to change the very pattern and basis of economic development. The TV industry is already well established in Africa. Location of TV industry activities in Africa, has certain favourable factors and advantages, in setting up this type of industry, as availability of acceptable level of education among the supervisory staff adequate technical and managerial skills developed over a long period, and availability of cheap labour are assured.

The organisation structure will vary from TV Station to TV Station in the industry and as such the pattern proposed herein can be considered only as suggestive and provisional.

A well-knit organisation structure headed by a Station Director and Manager, with the supporting staff will be developed progressively during project implementation. Soon after the plant becomes operative, a good number of project staff will be absorbed in the organisation. Certain additional staff also get added to ensure smooth and efficient management of the operating unit.

The Manager Production will have a degree in communication and will be accountable and answerable to the Station Director for all TV operations including planning, production, material management, TV utilities, quality control, production cost and budgetary control, TV safety, discipline and layout relations. His main function should be to ensure achievement for quality and quantity targets of production at reasonable cost and should constantly strive to improve TV performance. In the discharge of his multifarious duties and responsibilities he will be assisted by supervisors and adequate staff for day to day activities.

His duties are to ensure that targets are maintained through effective utilisation of 3 M's viz: Machinery, Men and Materials. These targets should be translated in terms of targets for individuals under him such as supervisors, skilled workers etc. He has to ensure that the equipment at his disposal gets prompt attention on breakdown is adhered to strictly. He has to further ensure that there is a strict quality control exercised over raw materials also. He will be assisted by Deputy Manager etc. who and will report to the General Manager.

GENERAL MANAGER (Comm.) He will report to the Station Director. His main duties will be prepare sales forecasts and budgets, study and monitor the export market for the company's products and advice on ways and means of increasing sales. He will also ensure that consumer complains are solved in an appropriate manner. He will also ensure that the right materials are available for production at the right time.

He will be responsible for all aspects connected with the export procedures and keep the management updated on all matters relating to the Govt. policies on polymers and Exports and above all the world market.

COMPANY SECRETARY & FINANCIAL CONTROLLER. He will report to the Station Director and will be responsible for the departments of administration, financial planning, budgetary control, cost accounting, tax management payroll accounting, etc. and all personnel matters. He will be assisted in their duties by their respective assistants to assist in day to day activities.

Manpower planning and production

The central issue here may be one of scale. Production is staffed bye personnel, but the process may be labour-intensive or capital-intensive. Either way, planning will include:

1.   Analysis of labour supply and demand factors in relation to skills and training needs.
2.   Procedure for manpower recruitment, together with selection processes.

3.   The formation of industrial relations policies necessary for effective work place bargaining, disciplinary measures and dismissal procedures.
4.   Analysis of the effective use of human resources.
5.   Conditions necessary to maintain adequate levels of motivation. The scale of the problem is likely to be directly proportional to the method of production.

The availability of main persons is not going to be easy since TV industry is currently in its infancy in Africa.

Training needs

The selection and training of the required manpower for the proposed project has to be planned in advance.

The key personnel should be selected and trained

suitably. The training would be carried out in the following manner:

Basic training on the concept of TV industry before construction begins with visits to similar TV Stations in other countries. On site training during the construction phase of the project. On job training during the commissioning phase of the project. On job training during operation of the TV immediately after commissioning. The training of the key personnel such as Station Director should be carried out in all the phases. The training of other operating personnel should be suitably carried out during the construction, erection and operation phases in addition to training them by visits to similar plants operating in other countries.

Besides training the key operating staff described above, in TV training should also be given to other employees at skilled operating level to enable them to understand the process equipment in the project and prepare them to operate a maintain their respective sections safely, efficiently and skilfully. The above training should be carried out during construction, commissioning and operating phases of the project.

Training is necessary in order to enable personnel to acquire the skills and knowledge necessary to perform a task to an acceptable standard. The length of the training period and training methods will, of course, vary from job to job. Training is essentially a learning process, and in order that progress can be successfully monitored certain conditions are necessary.

1.   The training needs of both the individual and the organisation shall be identified and analysed.

2.   Targets and standards shall be set for the trainee, which are within his capabilities.

3.   The pace of the training programme should reflect the trainee's ability to maintain progress in properly absorbing the same.

4.   The trainee shall receive regular feedback of results. Any problem areas shall be highlighted, discussed and resolved.

5.   As the trainee progresses the amount of information provided shall be gradually reduced, thus inducing a feeling of independence and competence.

It is common place to find a wide variety of tasks in an organisation and each will require varying degrees of skill, effort and responsibility. This being so, it is inevitable that rates of pay will also vary and the differentials between the jobs will reflect their relative values. However, other factors such as local market conditions, bargaining strengths and traditions also influence a company's payment structure and a great deal of planning is required if rationalisation is to be achieved. One technique which has been successfully adopted by many companies to establish an equitable wage structure is job evaluation. In a job evaluation exercise a comparison is made of common criteria over a range of jobs, and the resulting analysis may be linked to a points allocation or job ranking system, and hence to a wage scale.

In conducting a job evaluation exercise it is important to cover a reasonable variety of tasks within the whole spectrum. For each, a job description is prepared setting out details of the duties and responsibilities undertaken by the employee together with a statement about his working conditions. Very often this task is undertaken by work study personnel since, they are responsible for determining methods of operation and evaluating the work content of the job. Each job is assessed factor by factor, resulting in a comprehensive comparative analysis.

The individual is the most important resources of any company and only people who are well trained, well motivated and adequately rewarded will provide a positive and synergistic contribution towards the company's objective and its organisational health.

In most cases, the factors, which may be weighted

according to relative value, are as follows:

1.   Skill-education, experience and training.
2.   Effort-both physical and mental.
3.   Responsibility-for equipment, materials, initiative etc.
4.   Working conditions-general conditions, risk of accident and injury.

Pay policies affect not only individual employees but the whole organisation, and the rewards and objectives vary at different levels within the enterprise.

Industrial relations

An effective industrial relations policy is important, since is the system through which employees take part in decision-making and in many instances it affects the while atmosphere of employer/employee relationships. An industrial relations policy is essentially a set of rules whose determine procedures for negotiation on such matters as:

1.   Wage and salary scales.
2.   Terms and conditions of work.
3.   Disputes and grievances.
4.   Recruitment and dismissal.
5.   Other issues of mutual interest, e.g. closed shop, redundancies and joint consultation.

In order to promote an atmosphere of co-operation, and to minimise conflict, the needs of management and work people must be recognised by both sides. Trade unions exit to protect the interests of their members and improve their working conditions. Management, while aware of the pressures and constraints imposed by the trade unions, have a duty to maximise the use of resources at their disposal, which may be expressed in relation to profitability, return on investment, level of service, sales volume, market share and cost-effectiveness. The strategies adopted in attempting to solve industrial relations problems will vary from company to company, and indeed from union to union, but there is no doubt that they will be influenced by both internal and external factors.

Internal factors

1. The attitudes of employees to management, and management to employees.
2. The leadership style of management.
3. The bargaining strength of both parties.
4. The number of negotiating bodies.
5. The prosperity of the company.

External factors

1. The extent to which parent boards influence company management, and district officials

influence or control local shop stewards.

2.   Whether or not bargaining is conducted at plant, local or national level.

3.   Government policy towards industrial relations.

4.   The economic situations nationally, locally or within the company itself.

3.5   Describe the project implementation schedule.

Implementation of this project is a challenging task and calls for meticulous planning, scheduling and monitoring to realise the project goals within the budgeted cost and time frame. The goal can be achieved by adopting modern project management techniques.

To implement this project adequately, a team of engineers and project personnel having requisite education and experience are being appointed, to whom a detailed Work Breakdown Structure (WBS) in a logical order of activities, shall be supplied shortly, keeping in view cost estimation, scheduling, and to help monitor and control of the project. It is proposed to be formulated in conjunction with the objectives of each activity and goal settings. Project shall be programmed and controlled by network analysis techniques. Before the application of the network analysis techniques, the project personnel shall be acquainted with their capabilities in saving time, resources and costs. The training of project personnel and engineers at levels shall be provided for proper control of project progress and taking of timely corrective actions to re-align these efforts to meet ore stated objectives. It is proposed to gear programming and control system, i.e. project implementation system, which will ensure an integrated approach to project implementation. Project management activities shall be determined in advance and all activities carried out be project personnel as well as those to be contracted shall be identified.

Responsibility for project implementation shall be clearly defined. The forms of project organisation range from project oriented to functional organisation, while most of the cases are combinations of the two, with certain adaptations to prevailing conditions. It is impossible to over emphasise the importance of establishing a team of a task force for implementing the project with a designated leader to co-ordinate and guide its functions.

Project manager: Shall be responsible to the Board of Directors. The project manager shall be responsible for guiding and co-ordinating the efforts of all parties engaged in implementing the project, obtaining necessary government approvals on contracts. He is to control the project organisation with the promoters as well as with other agencies and organisations interested in the project. The manager shall have some staff to assist him, especially in checking expenditures to date and determining the present and future cost overrun or under run so that the project manager can take or propose to the Board pertinent corrective measures.

The network shall cover the pre-construction phase of the project indicating major administrative processes, since experience shows that some of them have frequently involved lengthy delays. In other words, it shall include the aggregate activities to be carried out be the principal parties participating in the implementation process.

The project budget shall then be prepared. Part or periodic payments to contractors which might be made at the end of certain intervals (e.g. weekly or monthly) throughout the time horizon of project implementation, shall be made by summing up activity costs per unit of time, which may be a week & month, and computing the cumulative cost at the end of each time interval. For the activities that are in process and are contacted or sub-contracted, the assumption of a linear time cost activity relationship shall be used for the sake of simplicity. In other words, expenditures are uniformly distributed throughout the duration of the activity.

PROJECT SCHEDULE

After an investment decision is taken, the main machinery and long delivery items must be ordered out at the earliest, forming the first major step in implementation of the project. It is foreseen that an engineering consultant will be appointed for carrying out the detailed engineering including basic engineering and procurement assistance to the client. It is also assumed that reputed and experienced contractors with adequate resources viz., men, materials, tools and tackles etc. will be engaged for execution of the construction and erection work. The purchase packages for auxiliaries shall be kept minimum so as to reduce the co-ordination efforts to the minimum. A great deal of co-ordination is required for constructing/erecting the new units. This task is feasible, provided the major activities of the project are co-ordinated and completed in the duration specified to achieve the respective milestones in time.

STRATEGY FOR TIMELY EXECUTION

It is important to deploy a team of experienced personnel for project execution and select the external agencies with due care for rendering the services and supply of equipment for the project. The project activities must be identified, planned and scheduled, and the progress monitored for timely project implementation. All the inputs to the project including financial resources must be identified and their inflow planned and arranged in time.

The project must be managed professionally with necessary co-ordination among the various agencies and requisite decisions taken promptly.

Establishment of an effective monitoring procedure for progress review and co-ordination.

In short, the following key factors would constitute the broad strategy for timely execution of all activities in a pre-determined manner as per schedule shown in the bar chart as to reach at a basis of regular production.

Early selection of an effective in-house technical team (TASK FORCE) by Government of Sao Tome, headed by a Project Manager for planning and executing the project.

i)   Proper choice of external agencies such as consultants for Project Engineering., Machinery suppliers, Construction Agencies etc. keeping in view their reputation/past performance and working experience in their fields.

ii)   Adequate use of computer-based PERT/ CPM techniques for project planning, scheduling and monitoring.

3.6   Description of the initial project Flowchart.

The project implementation phase embraces the period from the decision to start the project to the beginning of the commercial production. It includes a number of stages including negotiations and contracting, project design, construction and start-up.

3.7   Description of the requirement of Land.

For this project the Government of Sao Tome will participate with about 25.000 square meters of land. In selecting the land the following criteria should be followed.

1.   The land should be near to the main city of Sao Tome.
2.   The land should be properly connected by road.
3.   It should also have proximity to many end user clients.
4.   It should also meet various Government policy of
        achieving the social objectives.

2.08 Description of the requirement of Building.

The total requirement of building is as follows.

a. Adm. Building   400 Square Meters
b. Studios –3 (Three)   1500 Square Meters
c. Other Misc. Building   100 Square Meters

Total Square Meters: approximately 2000 Square Meters.
Cost for construction of about US$.750 per square meter (including air conditioning systems, primary electrical cabin etc.).

2.9   Description of some requirements for Plant & Machinery and their basis of selection.

The plant and machinery have been selected giving due consideration to the sophisticated nature of technology required. Detailed discussions were carried out with the foreign suppliers to ensure that the required capacities are practical with minimum capital and operating costs before the machinery were finally selected.

The machinery shall be selected from the most reputed foreign manufacturers of complete range of equipment. Keeping in view the following main factors:

a)   Past performance
b)   Existing machinery in Africa & abroad.
c)   Sales and service facilities in Africa

The main equipment is as follows.

STUDIO

1.   Post-production equipment – non-linear post-production, two ES7 stations.
2.   ENG – 2 DVCAM Camcorders
3.   Régie with 2 DXC-D30PK1 Digital Cameras - for cold light installation.
4.   Flexicart.
5.   Computer graphics suite.

Other technical data:
The whole installation is characterised by:
6. Informational web with UTP level, 5 cables.
7.   Phone system.
8.   Specific furniture.

Discrimination of Technical Data:

Studio
Cold lighting system.
2 Camera channels composed by:
1   DXC-D30PK Sony Digital Cameras 2 J18 Canon Zoom Lens
2 CA-537P Sony Camera Adapters

1   DXF-50 Sony Studio Viewfinder's
1   CCU-M3P Sony Control Camera Units 2 RM-
7P Sony remote CCU controls

2 50M Sony Multicore Camera Sets
2 Canon Focus and Zoom control sets
2 Intercommunication Headphones
2 Special tripods
1   charriot plane and curve
1   Teleprompter with 2 reading systems (monitors for individual speakers) Different microphones
with diverse directional characteristics
Audio components Video components

Video Régie

1 Recorder/Player BETACAM equipment
1 video digital mixer with 12 tracks
1 DVE equipment of digital effects for 3 channels
1 TBC Frame Synchroniser
1 Oscilloscope and Vectorscopy for control of cameras
1 Synchroniser Generators with Changeover
1 Intercommunication system with 4 places
1 Audio and Video Matrix with remote controls Video distributors

Audio Régie

1 audio mixer with 24 tracks

1 audio monitors
1 cd player
1 cassette deck
1   DAT
1   Mini-Disc
1   Level Detector for Audio Stereo Stereo Audio Distributors
Audio compressors Audio effects Microphones

Video Post-Production
2 Sony ES-7 on DVCAM/AVID hybrid non linear edition stations 1 Conventional BETACAM Edition Suite 2:1
Multi-track audio (8) on hard disc, Sound Scape type

3 ENG SETS:
Sony DSR-300 PK DVCAM Camcorder Digital Compact Report projectors
Various support materials (Batteries etc.)

3.10 Description of the Water Requirements for the project.

The requirements of water, separately for various matters are given in the following table.

Circulating   : Nil

Make-up   : Nil
Process   : Nil
Drinking   : 3,000 LPD

3.11 Description of the Steam requirement for the project.

A. Steam requirements and steam balance   : N.A.

B. Capacity and type of boiler with detailed specifications   : N.A.

C. Steam and energy diagram : N.A.

D.   Total energy generated / purchased (converted into M. K. : N.A. Cal) theoretical requirement of energy (in M. K. Cal)
at the various consumption stations and expected actual requirement at these stations.

E. If alternate processed are available, comparative energy   : N.A.
consumption figures for the various processes. If the
project is energy intensive, possibility of choosing alternate process in order to make the project less energy
intensive.

F. Steps proposed to be taken by the company to improve   : N.A.

energy losses efficiency and reduce energy losses (such

as power factor improvement, power load management,

optimising, illumination waste heat utilisation, etc.)

G. Scope for usage of solar / other renewable sources of   : N.A.
energy.

H. Any other measures contemplated in the direction of : N.A.
energy conservation and management.

3.123 Some information on Compressed air, fuel, etc.

Compressed Air

(a) Requirement   : N.A.
(b) Sources   : N.A.
(c) Arrangements proposed   : N.A.
(d) Cost at site with detailed calculations : N.A.

3.13   Details of the nature of atmospheric, soil and water pollution likely to be created by the project and the measures proposed for control of pollution. Indicate whether necessary permissions for the disposal of effluent have been obtained.

There shall not be any atmospheric pollution likely to be created by the project as there is no machine which has combustion resulting into air pollution or any chemical process which may release any gases which may result into an air pollution.

However, the use of DG set shall cause a small amount of air pollution. Considering the size of DG set, the pollution is within the permissible limits.

SAO TOME TELEVISION NETWORK

THE INDUSTRY

CHAPTER - 4

THE INDUSTRY

4.01 Description of the Television Industry in general.

The television industry can be described in the following broad categories.

1. Introduction

2.   Cable Networks

3.   DTH

4.   TV-media

5.   Earnings drivers

6.   Outlook

The detailed discussion in each of the categories is given below.

Introduction

The growing popularity of TV as a communication medium has resulted in the TV media sector undergoing a rapid transformation. From the black and white days of state controlled TV Station, to the highly colourful tunes of Channel V and MTV, the medium has certainly undergone a phenomenal change. Given its popularity, percentage ad spend has also increased proportionately on this medium.

Media pie (%)

	1995	1997
TV	62.5	68.8
Radio	20.9	15
Press	16.6	16.2

Source NRS

Entry of new channels post 1991

All over the world the telecasting has witnessed entry of new channels to cater to the various needs of world audiences. Channels have been launched in English as well as other regional languages. In many countries of the world till 1991, the state owned TV Station ruled the roost, as other players were not allowed to uplink and broadcast. However channels such as CNN, Star TV and BBC, which were offshore companies, could circumvent these regulations and telecast their programs into any country of the world. Cable operators then relayed the same and made it available to the common man through the cable television network.

Like many other countries, the State machinery controlled television. It was used as a propaganda tool for the party in power, with the opposition always at the receiving end. The customer had very little choice. The first steps towards more user choice began during the 1980s, which had to be telecast to a wider audience. TV Station used satellite channels for the telecast and the T.V. network was launched as an international channel.

The sports telecast by Channel 9 in 1985 and the Gulf War in the late eighties all played small but important cameos in educating the international viewer. With liberalisation in 1992 and crumbling tariff barriers televisions (read as colour TVs) became more easily available. The media revolution had started.

Major satellite channels avidly watched by viewers are Star TV, Sony TV, Home TV, BBC & CNN. There are other regional language channels which are major players in their respective territories.

Most of the channels that could not attain popularity rapidly suffered, since their advertisement earnings were not sustainable. The first round of the media wars is over. Management changes, i.e.

original promoters selling out to new management with deeper pockets, has become the order of the day. Alliances like the famous ESPN Star Sports arrangement also made headlines. Given the global trends of mergers and acquisitions, further consolidation is likely. Alliances and mergers make sense when the partners complement each other, like BBC and Discovery launched Animal Planet, CNBC and ABNI came together to launch a business channel called CNBC Asia.

Cable Networks

Antennas set up by either the end user or the cable operator receives the signals transmitted by the satellite. Local cable operators lay their own cables, set up control rooms, which can telecast 40 or more channels over a limited area. They charge the household a one- time connection charge of about US$.10 per point and a recurring monthly charge ranging between US$.1 to US$.5.

Initially, this was done in a very unorganised manner. The business required local knowledge and contacts, so every locality had its own cable operator. Collection was critical for the cable operator. For the end user, quality of telecast and a complete lack of standards became an issue. This lead to a shakeout and the formation of cable companies with money power which in turn tied-up with the local and small cable operators. Cable companies charge about US$1 per month to the local cable operators and support them with training and other infrastructure inputs. The business is immensely capital intensive and takes a long time to break even.

In many countries the operations of cable operators are regulated under the Cable TV Act which ensures that pornographic materials and other materials which are against culture and values or are detrimental to national interests do not get telecast. Recently this act has been amended to include foreign channels also.

Direct To Home

DTH is a new technology that circumvents the cable operators by directly delivering a bundle of channels to the end user. DTH involves transmission of encoded audio/ video signals (Ku band) via satellite. The end user needs an antenna to receive the signals and a decoder (set top box) to unscramble the encrypted signals. DTH services elsewhere in the world are Echostar and DirecTV (USA) and BskyB (Europe). Rupert Murdoch of Star TV fame owns BskyB.

The size of the antenna in DTH will be 1.5-2 ft in diameter, making it easy to install and transport. In conventional cable, since signals are in C band, an

8ft- diameter antenna is needed. The basic difference in the business model is the hardware costs in DTH. In a cable system, the user pays a one time connect fee and monthly rentals, while in DTH he has to invest in hardware.

The antenna will cost about US$200-300 and decoder will cost about US$200. The African viewer might be reluctant to incur such heavy installation costs. Quality of telecast in DTH is superior to Cable TV and viewer can receive up to 200 channels.

DTH will result in restructuring of the cable television industry. It will become imperative to have cash reserves to withstand the technology threat. Up gradation to fibre optic backbone will become necessary. A fibre optic network will cost about US$0.5mn per km as compared to US$0.1mn per km for coaxial cable. The stage is now ripe for consolidation.

TV Media features

In the Broadcasting business, it is only the industry leader who makes sizeable profits. The business is a game of asymmetrical payoffs. For instance, the top 5 channels account for 90% of ad spend.

Urbanisation and TV penetration is related. This may be due to the popularity of cable television that has resulted in increased colour TV sales. Rural penetration is low, although growing at a fast pace, because of dearth of specific program content to cater to that segment.

Liberalisation has resulted in the world viewer becoming more aware and conscious. This has resulted in the customer having more choice with the entry of a number of companies in different segments. Competition has resulted in companies increasing their marketing spend significantly.

Popularity of TV media is becoming higher. Increasing TV penetration leads to a reallocation of advertisement budgets with higher allocation for television at the cost of other medium.

TV channel operators use different business models to generate revenues. The critical component of any channel is the quality and type of programs they telecast. This determines their popularity, which in turn determines amount of advertisement revenues they can generate. They can do any one of the following:

Buy programming rights of program software from outside and collect advertisement revenue on their own. This model is followed by several TV companies, wherein they have a separate company in their fold, which develops all the content. The advantage is that re runs of serials/ programs become very profitable.

Selling time space to the producers for a fixed charge. Producers in turn are free to book advertisements at their own rates (there is an understanding on the time allocated for advertisement) and collect revenue. This is the basic model for many TV companies in which they sell prime time slots. The rights continue to be vested with the producer.

Earnings drives

The key factors that drive sector revenues are

Television penetration: Since the medium is television, increased television penetration will imply higher viewer ship. This will translate into higher advertisement spend allocation. This will also imply higher software production and demand for new programs.

Competition from other satellite channels would have an adverse impact on advertisement revenues, as advertisers have more choice in allocating ad budgets.

Government policies can have a big impact on the fortunes of the entire industry. When the DTH bill is passed in any country then, it will trigger a restructuring of the cable business.

Launching new channels targeted at specific segments, like regional channels within any country other areas having large pockets of ethnic population would lead to revenue growth. This will entail significant initial outlays.

Depreciation of the local currency would increase revenues as most of the program/ software companies export the programs overseas and payments are dollar denominated.

Advertisement revenue

As mentioned earlier, this is the primary source of income for TV channel operators. This revenue is directly co-related with the reach and viewer ship of a channel. Any channel's popularity depends on good quality programs, which is the software content. The business requires enormous initial investment in programs and revenues follow only with a time lag after the channel receives a minimum viewer acceptance.

Outlook

The sector has latent potential for growth on back of the exponential growth of cable TVs during the last 5 years. Television penetration in Africa is extremely low as compared to other developing countries like Malaysia, Pakistan, etc. in Asia. The number of channels has increased, implying higher demand for software programs.

Advertisement revenues, which are the barometer of channel popularity, will get dispersed over several competing channels. A shakeout is likely in both the channel and cable TV sectors. The biggest beneficiaries will be the content providers or the software houses. They will control the intellectual rights to the key element driving any channel's popularity.

Direct-to-Home, Digital Terrestrial Transmission and Conditional Access Cable Delivery have emerged as new delivery mechanisms. Breakthrough in technology would help open up avenues for these channels.

SAO TOME TELEVISION NETWORK

MARKETING

CHAPTER - 7

MARKETING

7.1   Description of the commercial viability of the project with regards to revenue generations.

Estimations shows that, after about 5 years of regular working, the CPLP Cultural Television Channel will be able to start exporting a great quantity of services and videos to Africa, Europe and South America. (See financial projections)

The economic feasibility of the project will occur with the commercialisation of several products, like:

1.   The sale and production of cultural programmes to various institutions like:
-       BBC
-       TV Culture Brazil
-       RTP
-       Discovery
Etc.

All over the world commercial networks for television cultural programmes have been created. Several television channels in different countries have regularly acquired programmes by producers spread out all over the world. Not only, several international organisations, like the World Bank and the United Nations, including FAO and Unesco, among many others, need programmes for public education, like programmes oriented to alert and to educate people concerning endemic and epidemic diseases. AIDS, malaria, tuberculosis are a very few examples.

These world institutions have made a great effort to develop television programmes with humanitarian objectives. But, in Africa – undoubtedly the continent most needed of such programmes – there is no television station, in present times, with capability to make face to such a need.

Therefore, the television teams that had been responsible for such programmes are, practically in its totality, placed in countries of the called First World, strange to the local population's concrete reality.

Programmes focusing new agricultural techniques or even oriented to agricultural, commercial and industrial education are essential elements of such a repressed demand.

The price for each institutional campaign, with an average of 10 films produced per campaign, is of about USD.15.000$ and the capacity of the CPLP Cultural Channel will be as follows:

First Group* - from 2 to 4 campaigns per month.
Second Group – from 1 campaign in two months to 1 campaign per month

This represents:
First Group   a potential annual income of about US$.540.000
Second Group a potential annual income of about US$.90.000

*First Group: Angola, Cape Verde, and Mozambique. Second Group: Guinea-Bissau, Sao Tome e Principe, East Timor (Lorosae). Reference Group: Brazil, Portugal.

2. The sale and production of no cultural programmes to other television channels, like:
-        series (novels)
-        talk-shows Etc.

Many countries of the region need to produce television programmes, but do not have capability to do it. Thus, they seem themselves obliged to search expensive productions in Europe. With only one programme sold-three daily hours-the channels of the First Group* would receive in incomes the equivalent of about two times of the whole investment. The Second Group countries will not have conditions at the beginning to product cultural programmes to other television channels.

*First Group: Angola, Cape Verde, and Mozambique. Second Grope: Guinea-Bissau, Sao Tome e Principe, East Timor (Lorosae). Reference Group: Brazil, Portugal.

The price for this type of television programme is:

Auditory programs   US$.12.500 per programme
Talk-shows   US$.4.000 per programme
Interviews   US$.4.000 per programme

The capacity of production, of these programmes by the countries of the First Group will be:

In the FIRST PHASE:
Talk-shows   2 per month
Interviews   4 per month
In the FINAL PHASE:
Auditory programmes 6 per month
Talk-shows   20 per month
Interviews   20 per month

Thus, the potential income will be:

FIRST PHASE
Talk-shows   US$.96.000 per year
Interviews   US$.192.000 per year

FINAL PHASE
Auditory programmes   US$.900.000 per year
Talk-shows   US$.960.000 per year
Interviews   US$.960.000 per year

The price of the novels is much higher. For each complete novel, with about 120 chapters, the price is about US$.2.500.000 and the capacity of the First Group of the CPLP Cultural Channel (in the final phase) will be of one novel per year.

3. There is the possibility of commercial use of the transmission time beyond the six hours reserved to culture and education.

The conventional day in television is of 18 hours, of which only 6 hours would be no commercial. We would have, therefore, 12 hours of no cultural transmission, which should be freely commercialised.

Even the period of six hours- divided into two sections of three hours, the first one dedicated to culture and second section of three hours to education-has a great potential for sponsoring.

Each commercial hour of transmission can include up to 12 minutes of advertising. The price for each 30 seconds of advertising is:

Noble time   from 6PM to 11PM   a b o u t
US$.400 each 30"
Normal time   rest   about US$.200 each 30"

Concentrating the commercial programming in the noble time. With 4 hours of transmission in this period, we would have:

Educational television: from 7AM to 9AM
From 3PM to 4PM
Cultural television: from 4PM to 7PM
Commercial television: from 9AM to 3PM
From 7PM to 1AM

Thus, the commercial period would comprehend 6 hours in normal time, 4 hours in noble time and 2 hours in normal time again.

Begin 12 minutes per hour for advertising, the commercial period could have:
Noble time   48 minutes   24   films   US$.9.600 per day
Normal time   96 minutes   48   films   US$.9.600 per day
Total of the potential income........... US$.19.200

per day

> or US$.576.000 per month
> or US$.6.912.000 per year*

It is believed that in the first phase of the project (after one year), the CPLP Cultural Television Channels of the First Group will be able to start with an income from commercial advertisement of about US$.1.000.000 per year-value which is predicted to increase in the follow months.

Obs. * The amounts for Sao Tome are about US$.3.500.000 per year

4. The rent of the studios to television teams of other countries can be another source of incomes. Many producers who cover events in Africa need to move many times to their countries of origin during the video works, because there is not technical support in Africa. The same phenomenon happens with the cinematography and the journalism productions. It is not difficult to imagine, for example, the serious problems journalism teams have had, for example, with essential components like batteries, lighting, electronic components etc., which only can be easily find in Europe.

The period of renting of a television studio is of 12 hours. Each period has price of about US.4.000. After the first year, the studio should be rented for 12 hours per each period of 3 days. So, the rent per month, in this period (First Phase), will be able to generate incomes of about US$.120.000 per month in the First Group.

5. Support services to other television networks in all areas, including novels, mini series, docdramas series etc.

Many countries, principally in Africa, do not have technical conditions to develop this kind of programmes, but they have a strong internal repressed demand in this sense.

Each team, abroad, has a price of US$.1.000 per day (two people), after the costs of dislocation, hotel and meals. In the first months (first phase) it will be organised one team for works in other countries. In the final phase it is predicted to have up to 5 teams with such function. The capacity of each team is of about 20 days per month. Therefore, the support to other television networks-when took in its full potentiality-will be able to represent up to 20 days per month in the first phase. This represents a potential income of about US$.240.000 per year, and 100 days per month of works in the final phase, signifying about US$.1.200.000 per year of incomes, always referring to the First Group countries.

6. Colloquies, seminars, videoconferences and meetings of different natures.

The building of the CPLP Cultural Television Network will have a medium size auditorium, (First Group), with all conditions to receive seminars, colloquies and meetings of the most diverse nature. Such seminars, videoconferences etc. are important not only for the increase of the incomes, but also to attract specialists of the most different areas, being an important element for the development of all region as well as for the diffusion of the local, regional and continental cultural values.

The meetings, seminars and colloquies attract, in average, about 300 people per event. The price-excepting meals, hotels and transportation-per each participant is of about US$.50 per day. It will be a capacity for up to four events of this type per month, what could represent an income of about US$.60.000 per month or US$.720.000 per year (final phase-after 5 to 7 years).

The project should also turn possible.
-    classes of professional formation in diverse disciplines
-    support for a multimedia high technology centre (First Group)
-    support for a multimedia high cultural centre (Final Group)

SAO TOME TELEVISION NETWORK

PROFITABILITY & CASH FLOW

CHAPTER - 8

PROFITABILITY AND CASH FLOW

8.1   Estimation of cost of production and working results for the first five years of operation.

The estimated cost of production of working results for the first five years of operation are given in the chapter "Financial Projection".

8.2   Cash flow statement for the company as a whole, for five operating years of the project based on the estimates of working results.

A detailed cash flow statement for the company as a whole for five operating years is given in the chapter of "Financial Projection".

8.3   Projected balance sheet for five operating years for the company as a whole.

The balance sheet for five operation years for the company as a whole is given in the chapter of "Financial Projection".

SAO TOME TELEVISION NETWORK

ASSUMPTIONS

CHAPTER - 9

ASSUMPTIONS

1.   The Sao Tome Television Network shall generate the income from the following sources.

a)   Sale of cultural programmes.

b)   Sale of commercial programmes.

c)   Sale of advertisement time during transmission of TV programmes

d)   Hiring of studios.

e)   Supply of technical services

f)   Hiring of conference hall

The detailed assumptions for each of the above mentioned activities are as follows:

a)   Sale of cultural programs:

I.   It is assumed that four cultural programs shall be produced per month in the first year, which can be sold to other countries. This
figure will increase to 5 programs per month in the second year, six programs per month in the third year and so on.

II.   It is assumed that the selling price shall be US$.75 per program.

b)   Sale of commercial programs:

I.   The commercial programs consist of talk shows, interviews and auditory programs.

II.   It is assumed that two talk shows shall be produced per month in the first year, three talk shows shall be produced in the third year, four talk shows per month shall be produced in the third year and
so on.

III.   It is assumed that the selling price of one talk show program shall
be US$.2000.

IV.   It is assumed that four interviews shall be produced per month in the first year, five interviews shall be produced in the third year, six interviews per month shall be produced in the third year and so on.

V.   It is assumed that the selling price of one interviews program shall be US$.2000.

VI.   It is assumed that no auditory programs shall be produced in the first and the second year. Only when the people have two years of experience then in the third year one auditory program shall be produced per month. In the fourth year two programs shall be produced per month and so on.

VII.   It is assumed that the selling price of one auditory program shall be UD$.6.250.

c) Sale of advertisement time during the transmission of TV programs:

I.   There shall be a total transmission of 8 hours per day in the first year. It will go on increasing by two hours in the second and subsequent years.

II.   Out of which advertisement shall be available for programs with transmission period of four hours in the first year. The programs in
which advertisements could be available shall increase by two hours per year in the second and

subsequent years.

III.   In one hour T.V. transmission, 12 minutes of advertisement shall be allowed.

IV.   Out of total T.V. time 33 % time shall be considered as prime time and remaining 67% shall be considered as non prime time, in the first, second and third years. From the fourth year onwards the prime-time advertisements shall remain constant.

V.   The advertisement rates shall be US$ 400 for one minute of prime time advertisement and US$ 175 for one minute of non prime time advertisement.

d)   Hiring of studios:

I.   It is assumed that the studios shall be taken on hire for a shift of 12 hours, two times in a month in the first year. It will increase to three times in a month in the second year.

II.   The rent per day shall be US$.2.000.

e)   Supply of technical services:-

I.   It is assumed that one team, consisting of two technically qualified people shall be available in the first year. In the second year two such teams shall be available. In the third year three such teams shall be available and so on.

II.   It is assumed that in the first year the team

shall be hired for 10 days in a month. In the second year the teams shall be hired for 12

days in a month. In the third year the teams shall be hired for 14 days in a month and so on.

III.   The rate of hire shall be US$.500 per team per day.

f)   Hiring of conference hall :-

I.   It is assumed that the conference hall shall be taken on hire for two times in a month in the first year, three times in a month in the second year and four times in a month in the third year and so on.

II.   The hire charges shall be US$.250 per conference.

2.   The unit is planning transmission of three hours of educational programs, three hours of cultural programs and two hours of commercial programs per day in the first year. Thus, there shall be a total transmission of eight hours of programs per day in the first year. The total transmission hours shall increase by two hours per day in the second and subsequent years. Thus, in the second year there shall be total transmission of 10 hours per day and in the third year there shall be a total transmission of 12 hours per day. The cost of software production is assumed at US$.1.250 per hour.

3.   The cost of miscellaneous consumable

items is assumed at US$.18.000 in the first year. This will increase by 10% in the second and subsequent years.

4.   The cost of stores and spares consumed is assumed at US$.21.000 in the first year. It will increase by 10% in the second and subsequent years.

5.   The cost of repairs and maintenance is assumed at 0.1% of the total cost of plant and machinery in the first year. This cost will increase up to 0,3% in the fifth year.

6.   It is assumed that the unit shall use 1.000 units of power every day or 365.000 units of power in the first year. The consumption of power shall increase by 10% in the second and subsequent years. The cost of power is assumed at 2.5 cents for one unit.

7.   It is assumed that the unit shall consume 1.000 litters of diesel per month. The cost is assumed at 16.5 cents for one litter. It will increase by 10% in second and subsequent years.

8.   The unit shall have a technical staff of 35 people and administrative staff of 15 people.

9.   Depreciation is calculated on reducing bal-

ance method. The rate of depreciation is taken at 10% for building, 25% for plant and machinery, 10% for furniture and fixtures and 20% on vehicles.

10.   Advertisement expenditure is assumed at 0.2% of the sales.

11.   It is assumed that the stock of raw material and stores shall be 30 days of consumption, receivables shall be 45 days sales and creditors shall be 15 days purchases.

12.   The term loan shall be for a total period of 25 years. It shall have a moratorium of five years. It shall be repayable in 20 equal yearly instalments.

PROJECT REPORT

OF

TIMOR LOROSAE CULTURAL TELEVISION NETWORK

AT

TIMOR LOROSAE

ASIA

PROMOTERS

REPUBLIC OF TIMOR LOROSAE

ASIA

TIMOR LOROSAE TELEVISION NETWORK

INTRODUCTION

CHAPTER - 1

INTRODUCTION

1.1   Description of the organisation through which this project is being taken up.

The CPLP is an international organisation, created in 17 of July of 1996, with Headquarters in Lisbon and consisting of the following State Members: Angola, Brazil, Cape-Verde, Guinea-Bissau, Mozambique, Portugal, Sao Tome e Principe and East Timor (Timor Lorosae).

The CPLP has as its objective the politic-diplomatic relationships between its Members, mainly referring to the United Nations and the World Bank, the relationship EU/MERCOSUL and the accomplishment of the Europe-Africa Summit.

The CPLP is also dedicated to the co-operation, particularly in the economic, social, cultural, legal and technical-scientific matters; as well as to the projects of promotion and broadcasting of the Portuguese Language, nominated to the improvement of the International Institute of the Portuguese Language and the creation of a Bibliographical Fund.

All its decisions are taken by consensus.

2.2   Description of the main reasons for setting up an African and Asian Cultural Television Network.

Africa is a continent submerged into dramatic problems. All these humanitarian problems have in the education and culture their most direct and objective root.

The African countries speaking Portuguese are among the African countries with the lowest level of development.

However Timor Lorosae is not located in Africa, but in Asia, it is member of the CPLP and its situation, given the recent political events, puts the country in the front line for the Cultural Television Network Project.

The whole situation is so dramatic that the implementation of the CPLP Cultural Television Network is more than urgent.

It is hoped that with this Television Network the

people shall have better understanding of various diseases and other problems of life. e.g. AIDS is prevalent disease in Africa. But as the people are illiterate the information about the disease cannot be spread by any other method. In such cases the Television Network becomes a very effective audio-visual means of communication.

The project also intends to be a base of a massive production of health campaigns, specially to rein-force the combat against the AIDS, the infect-contagious diseases- endemic or epidemic-through a process of global education of the local and region-al populations.

The CPLP Cultural Television Network is a project for peace and its objective is not to compete with local television channels already installed, but to interchange with them, promoting a future global web dedicated to knowledge and development.

1.03 Description of the main objects of the project.

The major objective of the project is to assist the CPLP countries to develop the first African televi-sion network channel oriented to cultural and edu-cational themes with the follow orientation:

-       diffusion of the Portuguese language
-       education in all areas

-        reinforcement of the local and regional cultures
-        formation of technical personal on communication and tele-education

-        formation and information for a better participation of the CPLP nations in the process of global development
-        to support the implantation of the Institute of Portuguese Language

The development of such a project will create an irradiation pole from the countries with the project implementation to the whole African continent, constituting an African network of research and culture, discovering and valorising different cultural aspects and, consequently, reinforcing identity factors.

With the stronger cultural identity, it is expected to have lower levels of violence in social behaviour as well as a relevant increase of the educational performance in all its sectors.

Systematic educational objectives are complementary to the general orientation of this project, which is cultural in its fundament.
A very large positive impact is also to be expected with respect to several sectors of the economy, greatly improving the overall quality of services.

1.4   Description of the mission status of the project.

The mission of the project is:

-    to rescue lost values and cultural information
-    to reinforce historical elements
-    to rescue old artworks
-    to give a new value to new artworks
-    to communicate practical and objective information on health
-    to communicate practical and objective information on agriculture
-    to communicate practical and objective information on economy

1.05 Description of the philosophy of the project.

The CPLP Cultural Television Network will have in its philosophy, as a support to its mission, the follow fundaments:

-    to work with local people
-    be focused on quality
-    intercommunication

Three words can define the CPLP Cultural Television Network philosophy:

People, quality and intercommunication

Following these three key words of its philosophy, a single sentence also defines its mission: "A Continent of Culture".

Two main sectors characterise its global structure:
1. Cultural Sector
1.1.   To develop programmes on:
-        Music
-        Architecture (also vernacular)
-        Archaeology
-        Literature and poetry
-        History
-        Tourism
-        Dance
-        Plastic arts
-        Astronomy
-        Science in general
-        gastronomy
-        health
-        anthropology
-        local cultures
-        others

1.2   The programmes can be of all types, the most important is to be attentive to communication. All programmes must reach the highest possible number of spectators.

2. Educational sector.

2.1   Systematic education in short practical courses.

2.2   Educational series with practical and objective information on:

-       health care
-       family
-       agriculture
-       managing
-       services
-       civil construction

As the CPLP Cultural Television Network must not be supported by the respective governments in the future, it will have a daily time of free commercial programmes up to 12 hours. This is a good solution to pay its costs and support its development.

1.06 Description the magnitude of the project.

The project proposes a Television transmission time of eight hours per day in the first year. It will increase by two hours per day in the second and subsequent years.

The project also envisages training of local people in all the related areas of Television. It is proposed to develop a local pool of talent which can be utilised not only in this particular country but also in other countries of Africa.

The total cost of the project of the Timor Lorosae

Cultural Television Network is US$.10.600.000. Out of this it is proposed to obtain a soft loan of US$.8.800.000. The remaining amount of US$.1.800.000 shall be contributed by the Government of Timor Lorosae in terrain for the installation of the respective building, infrastructures and logistic support.

For the purpose of the Timor Lorosae Cultural Television Network a company shall be formed with the participation of the Government. The Contribution of the Government in the form of share capital shall be US$.1.800.000.

The term loan shall be on soft terms. The total period of the loan shall be 23 years. Out of which five years shall be a period of moratorium. The entire loan shall be repaid in the remaining 17 years, in 17 equal yearly instalments.

The Government of Timor Lorosae will give the sovereign guarantee for the referred loan.

TIMOR LOROSAE TELEVISION NETWORK

THE COUNTRY

CHAPTER - 2

THE COUNTRY

2.1 - Description the level of development of Timor Lorosae.

East Timor, occupying the eastern part of the island of Timor, is situated between Australia and Indonesia.

Following a referendum in August 1999, East Timor was placed under the supervision of the United Nations. On 9 December 1999, a Trust Fund for East Timor (TFET) was established with the assistance of funding agencies. The TFET grant projects are formulated and supervised by ADB and the World Bank.

The economy was severely affected by the violence and destruction that followed the referendum, with some estimates indicating a 40 percent decline in real gross domestic product (GDP) in 1999. Public and private property was damaged, the agricultural cycle and trade were disrupted, and a substantial part of infrastructure was destroyed. Low cash activities; low demand; and lack of capital, infrastructure, and a transport system constrain the economy. Unemployment is high and humanitarian assistance continues to be an important source of support for a great part of the population.

As conditions in the country gradually returned to normal, the economy rebounded. Given the low base, real GDP growth was estimated at 15 percent in 2000, largely as a result of reconstruction work and the gradual restoration of commerce, business, and basic services. Because of favourable weather, the output of maize, rice, and coffee (the major export commodity) also improved in 2000. Inflation slowed, along with reduced regional disparities in staple food prices.

The recovery process is expected to gather momentum and be led by agriculture, construction, and services; prices are expected to stabilise in the coming years. The combined sources budget of East Timor for FY2001 (ending 30 June) provides for a total revenue of $25.1 million and total expenditure, including net lending, of $183.0 million. The overall deficit of $157.9 million (about 55 percent of GDP) will be financed largely from external grants.

In 2000, the National Consultative Council adopted the US dollar as the legal tender, although the Indonesian rupiah continued to be used. The Central Payments

Office began providing basic depository and payment services. Several regulations related to banking and the payments system were enacted. The Central Fiscal Authority has also started revenue and budgetary functions. The Department of Economic Affairs and the National Planning and Development Agency were created.

The intense level of violence that followed the 30 August referendum resulted in a massive population displacement, both within and across the borders of East Timor. Food distribution and marketing systems, together with commerce and essential services, were paralysed and the agricultural cycle was disrupted. In addition, although infrastructure and property damage was extensive, agricultural damage was less severe as crops had already been harvested and only limited burning of the relatively minor second season crop in fields, occurred.

The main damage was to livestock, food and seed stocks, which were either looted or burnt, and to disruption in the agricultural cycle, which either delayed or will delay key operations such as land preparation and planting. The main constraint to planting the main crop of maize, in November/December, is population displacement, which will mean that in some areas a relatively large proportion of farmers will not have returned to complete sowing. In addition, although the availability of maize seeds was earlier envisaged as a serious impediment, international efforts to distribute seeds, through the

UN and NGOs, have been highly effective, covering a large proportion of requirements.

As the remainder will come from farmer exchanges and use of stored seeds that were not destroyed, the loss in production in 2000 will be lower than thought earlier. As rice planting/transplanting does not need to be completed till mid January, there is greater scope for recovery. Population displacement and seed shortages are not considered to be major problems for the current rice season, though there will be some reduction in area planted due to the marked reduction in animal and mechanised traction, which are essential for land preparation. Although the rainy season started somewhat late, rainfall during November/early December was generally favourable, benefiting crops.

Although the level of disruption in agriculture was less than anticipated at the height of the crisis, there will inevitably be a notable decline in cereal production, given the constraints outlined above.

East Timor is known to have pockets of malnutrition but it is not well documented. The El Niño-related drought, which seriously affected food production in 1998, and the financial crisis in Asia, which resulted in tremendous price hikes, greatly exacerbated problems of food availability over the last year.

East Timor is the eastern half (approximately) of the

island of Timor. It is a small mountainous territory, around 14 500 km2, in the Indian Ocean some 500 km to the north of Australia. Even before recent events and population displacement, East Timor was among the poorer regions in Asia. Approximately 50 percent of the population are considered below the poverty line, life expectancy is around 56 years, whilst only two out five people are literate. The level of infrastructure and the provision of essential services are also poor, with only 30 percent of households having access to potable water and 22 percent having electricity. Less than half (49 percent) of the number of villages were accessible by paved road before the crisis.

Although the environment and the natural resource base of the island are extremely important in development they are not as conducive as in many other small island states. The economy is primarily agricultural, contributing the largest share to GDP, employing almost three quarters of the workforce, providing over 70 percent of the population with their main sources of livelihood and offering the greatest potential for exports and trade. The sector, in turn, is largely dominated by subsistence production of primary staples, (maize, rice, cassava and sweet potatoes), employing family labour and little purchased inputs.

Land use, however, is highly constrained by rugged topography, poor soils, erosion and unpredictable rainfall. In general, soils do not support heavy vegetation, which limits the scope for crop diversification. Because of these constraints, producers are largely risk averse, employing a strategy aimed at minimising potential losses rather than maximising yields. Periodically the island is affected by El Niño related weather anomalies, which further reduce food production. It is estimated that around 600 000 hectares are suitable for agriculture of which approximately half are currently being used.

In addition around 200 000 hectares are suitable for rearing livestock, which, hitherto, were the territory's second largest export after coffee. However, after the recent events and given security, ownership concerns and lack of capital investment, the sector remains well below potential.

Although fisheries offer considerable scope, a very small proportion of potential is currently exploited as the industry remains largely under-developed. Less than half the work force in fisheries is full time, and the equipment used inefficient. The main fishing region is off the coast around the capital Dili and Atauro.

The potential for developing forestry is limited due to severe deforestation mainly connected with slash and burn agricultural practices. In general there is little scope for further development, as only small tracts are suitable for expansion. Hitherto, sandalwood was an important and valuable export but over-exploitation led to the cessation of exports in 1930s.

The climate is dominated by intense north-east monsoon rainfall from November to March, followed by a pronounced dry season. In northern parts of the island, this constitutes the main wet season, whilst in the south, the rainy season is more prolonged spanning December to June, increasing the potential for secondary crop production of maize and possibly rice. The northern coast has the lowest average rainfall, ranging from 600mm to 1 000 mm per annum. In contrast, total precipitation is highest in central mountainous areas with a range from 1 800 to 3 000mm. Rainfall in southern coastal areas ranges from around 1 250 to 3 000mm. See figure 1, for an indication of rainfall distribution.

On 5 May 1999, agreement was reached between the Governments of Indonesia and Portugal for phased transition in the territory. The three phases agreed encompassed: (1) the period between 5 May 1999 and 30 August 1999 for preparation and execution of a referendum on independence (2) the post referendum phase, during which Indonesia would retain responsibility for security and Government services and (3) the transition to independence overseen by a UN-led transitional civil and peace keeping authority. Contrary to the agreement aimed at peaceful transition, however, widespread civil unrest and violence followed the August 30 referendum, in which the majority (78 percent) of Timorese rejected autonomy in favour of independence.

Although precise figures are still unknown, many people were killed in the violence, whilst hundreds of thousands were internally displaced or exiled as refugees to West Timor. In addition, enormous damage was inflicted on infrastructure, essential services and property. Available estimates, for example, indicate that Dili, the capital, and smaller urban centres were extensively damaged with up to 50 percent of social infrastructure and buildings destroyed. Moreover, individual household assets (primarily livestock and basic goods), were also destroyed or looted. Overall, therefore, in the intense wave of post referendum violence, essential fabrics of the economy and commerce were almost entirely demolished.

Although the situation has improved significantly following Security Council Resolution 1264 and the deployment of an international peace keeping force (INTERFET), the reconstruction of East Timor will require considerable time and resources, both capital and human. In the interim, humanitarian assistance will be vital to safeguard food security and the provision of basic needs.

In addition to agricultural and property damage,

East Timor lost essential human resources as edu-
cated leaders and service staff were specifically
targeted in the violence, while the majority of civil
servants left the province as many were Indonesian. The gap in administration, service and skilled
personnel will invariably have an adverse affect on
economic functions in the short to medium term.

TIMOR LOROSAE TELEVISION NETWORK

THE PROJECT

CHAPTER - 3

THE PROJECT

3.1   Description of the initial steps to be taken
for the implementation of the project.

The first objective of the CPLP Cultural Television

Network is to install its structures in the following countries:

- Sao Tome e Principe
- Cape Verde
- Mozambique
- Guinea-Bissau
- Angola
- Timor

Thus, these six countries will structure the basic network of the whole project.

Angola, Mozambique and Cape Verde will form a FIRST GROUP; Sao Tome e Principe, Guinea-Bissau and East Timor will form a SECOND GROUP; and, finally, Brazil and Portugal will form a called REFERENCE GROUP.

However, the project will start at the same time in all countries. A more precise description of the meaning of each group is made below.

The basic physical components of the project are:

1.   The creation of companies, in each of the States of the CPLP, oriented to the objectives above referred.
2.   The State of each country – who will be the responsible and guarantee of the payment of the financing – will also have a participation in the com-

panies.

3.   All strategies will be oriented to lowest costs in long term.

4.   The whole project will be developed by phases.

5.   The timetable of transmission will be defined case by case.

6.   The project will include:

6.1.   a complete plan for an equipment network.

6.2.   a complete plan for a software network.

6.3.   all procurements

6.4.   equipment transportation and installation

6.5.   all configurations

6.6.   creation of the infographic system

6.7.   technical formation of the personal

6.8.   architecture and engineer projects for the building

6.9.   construction of the building organised in phases

1.2   Description of the procedure by which the necessary programmes will be produced for transmission on the channel.

The television programs are known as software in this field. The unit is planning transmission of three hours of educational programs, three hours of cultural programs and two hours of commercial programs per day in the first year. Thus, there shall be a total transmission of eight hours of programs per day in the first year. The total transmission hours shall increase by two hours per day in the second and subsequent years. Thus, in the second year there shall be total transmission of 10 hours per day and in the third year there shall be a total transmission of 12 hours per day. The cost of software production is assumed at US$.2.500 per hour. The amount of US$.2.500 per hour is taken based on cost of production of such programmes in Portugal.

To contract a similar system from other television channel, only for one country, for a transmission period of 6 hours daily, the cost in, market (eg. Portugal*) is of about US$.500.000 per month. This cost represents a total of about US$.6.000.000 per year. In 16 years, excluding the first year of implementation, the costs for contract the six daily hours of transmission from other television channel, in present market price, is of about US$.100.000.000. The six countries would have an estimated cost of about US$.600.000.000 in a 16 years period of time.

These costs are based on the prices presently used in Portugal, as for example the programme Hora Viva – Seguranca em Directo, transmitted every day, excluding weekends, from 7 to 10 AM.

Another example is the channel Canal Noticias Lisboa CNL. Only in its first year of installation, it was expended about US$.10.000.000, and the equipment as well as the building were not property of the channel, but rented.

In fact, however the project will be implemented by phases, after 5 to 10 years of regular work, it is predicted to have reached a similar value in incomes as the examples above.

*value took in the Portuguese market, below the average of the European prices in about 30%.

The first ten months of works will be oriented to:
-        design of the equipment and software structures
-        start the first productions
-        elaboration of the projects of architecture and engineering
-        procurement and acquisition of all equipment
-        construction and supervision of the works
-        installation and configuration of the equipment

So, the first year must be used to the implantation of the project. However, as to accelerate the chronogram, some operations should be started before this timing, temporarily installed in a different building.

The first phase will be oriented to:

- video productions (external and internal)
- edition
- elaboration of cost programmes

3.3 Description of the requirements of Power, Fuel and Water. POWER

It is assumed that the unit shall use one thousand units of power of everyday. Thus, the unit shall use 365.000 units of power in a year. The costs of power is assumed at two cents for one unit. The consumption of power shall increase by 10% every year.

FUEL

The unit shall use D.G. set for generation of electricity, when there is an interruption in the supply. The hours of disruption in the normal supply can not be predicted accurately. Hence, the hours for which the D.G. sets shall be used also can not be predicted accurately. However, it is assumed that the unit shall utilise one thousand litters of diesel per month. The cost is assumed in about 13 cents for one litter. Hence, the total cost for one thousand litters shall be US$.130. The consumption of diesel shall increase by 10% every year.

WATER

The unit does not need water for fixed commercial

operation. However, it shall have a strength of 50 people as employees. Water shall be required for drinking and sanitation purposes. It is assumed that 3000 litters of water shall be required everyday. The cost of obtaining this water is assumed at US$.4 everyday. The consumption of water shall increase by 10% every year.

3.4   Description of the manpower requirement of the project and description of the steps taken by the company to train the manpower.

The total staff strength shall be 17 people in the first year.

The system of modern TV Station is designed to achieve sustained operating efficiency and trans-mission. This, of course, entails a certain degree of sophistication in the production and transmission. However, considering the socio-economic situation in Timor Lorosae, a reasonable balance has to be struck to obtain optimum performance and at the same time create gainful employment. While work-ing out the manpower requirement for this project to be kept on direct rolls of the company totalling 50 staff and operators, the above consideration have been kept in mind.

Manpower requirement

The direct manpower required for the proposed Unit is about US$.720.000 per year. The manpower requirement as indicated in this chapter has been planned keeping in view the following guide-lines:

Effective co-ordination among the various departments. Judicious distribution of responsibilities.

Capacity utilisation of the TV Station with optimum manpower.

Details of manpower requirement for the TV Station is given in the financial section.

In line with the prevailing practice, the Security guards, office peons and unskilled labour etc., are normally employed on contractual/casual basis, and their cost has to be included in the Factory overheads. However in this report provision has been made to employ the above staff and their salaried have been included.

The various departments proposed shall be under the direct responsibility of the Station Director. He shall be assisted by Deputy Directors, who will look after the complete TV Station and its various day to day activities and Marketing. They shall be responsible for achieving the envisaged targets and sales forecasts. They shall be assisted by a team of Managers from production, marketing and accounts.

When evaluating an investment project from an employment point of view, its impact on both unskilled and skilled labour has been taken into account. Not only direct employment, but also indirect employment has been considered. Direct employment refers to the new employment opportunities created within the project; indirect employment concerns job opportunities created in other projects linked with the project which is being formulated.

The implementation of large and sophisticated projects generally contributes to the development of local skills and capabilities in a country. Furthermore, they help to change traditional values, attitudes and the behaviour of the society, to build up an enterprising spirit among the people, to develop a desire for changing and improving the existing conditions of life, to introduce better work discipline and thus to change the very pattern and basis of economic development. The TV industry is already well established in Africa. Location of TV industry activities in Africa, has certain favourable factors and advantages, in setting up this type of industry, as availability of acceptable level of education among the supervisory staff adequate technical and managerial skills developed over a long period, and availability of cheap labour are assured.

The organisation structure will vary from TV Station to TV Station in the industry and as such the pattern proposed herein can be considered only as suggestive and provisional.

A well-knit organisation structure headed by a Station Director and Manager, with the supporting staff will be developed progressively during project implementation. Soon after the plant becomes operative, a good number of project staff will be absorbed in the organisation. Certain additional staff also get added to ensure smooth and efficient management of the operating unit.

The Manager Production will have a degree in communication and will be accountable and answerable to the Station Director for all TV operations including planning, production, material management, TV utilities, quality control, production cost and budgetary control, TV safety, discipline and layout relations. His main function should be to ensure achievement for quality and quantity targets of production at reasonable cost and should constantly strive to improve TV performance. In the discharge of his multifarious duties and responsibilities he will be assisted by supervisors and adequate staff for day to day activities.

His duties are to ensure that targets are maintained through effective utilisation of 3 M's viz: Machinery, Men and Materials. These targets should be translated in terms of targets for individuals under him such as supervisors, skilled workers etc. He has to ensure that the equipment at his disposal gets prompt attention on breakdown is adhered to strictly. He has to further ensure that there is a strict quality control exercised over raw materials also. He will be assisted by Deputy Manager etc. who and will report to the General Manager.

GENERAL MANAGER (Comm.) He will report to the Station Director. His main duties will be prepare sales forecasts and budgets, study and monitor the export market for the company's products and advice on ways and means of increasing sales. He will also ensure that consumer complains are solved in an appropriate manner. He will also ensure that the right materials are available for production at the right time.

He will be responsible for all aspects connected with the export procedures and keep the management updated on all matters relating to the Govt. policies on polymers and Exports and above all the world market.

COMPANY SECRETARY & FINANCIAL CONTROLLER. He will report to the Station Director and will be responsible for the departments of administration, financial planning, budgetary control, cost accounting, tax management payroll accounting, etc. and all personnel matters. He will be assisted in their duties by their respective assistants to assist in day to day activities.

Manpower planning and production

The central issue here may be one of scale. Production is staffed bye personnel, but the process may be labour-intensive or capital-intensive. Either way, planning will include:

1. Analysis of labour supply and demand factors in relation to skills and training needs.
2. Procedure for manpower recruitment, together with selection processes.
3. The formation of industrial relations policies necessary for effective work place bargaining, disciplinary measures and dismissal procedures.
4. Analysis of the effective use of human resources.
5. Conditions necessary to maintain adequate levels of motivation. The scale of the problem is likely to be directly proportional to the method of production.

The availability of main persons is not going to be easy since TV industry is currently in its infancy in Africa.

Training needs

The selection and training of the required manpower for the proposed project has to be planned in advance.

The key personnel should be selected and trained suitably. The training would be carried out in the following manner:

Basic training on the concept of TV industry before construction begins with visits to similar TV Stations in other countries. On site training during the construction phase of the project. On job training during the commissioning phase of the project. On job training during operation of the TV immediately after commissioning. The training of the key personnel such as Station Director should be carried out in all the phases. The training of other operating personnel should be suitably carried out during the construction, erection and operation phases in addition to training them by visits to similar plants operating in other countries.

Besides training the key operating staff described above, in TV training should also be given to other employees at skilled operating level to enable them to understand the process equipment in the project and prepare them to operate a maintain their respective sections safely, efficiently and skilfully. The above training should be carried out during construction, commissioning and operating phases of the project.

Training is necessary in order to enable personnel to acquire the skills and knowledge necessary to perform a task to an acceptable standard. The length of the training period and training methods will, of course, vary from job to job. Training is essentially a learning process, and in order that progress can be successfully monitored certain conditions are necessary.

1.   The training needs of both the individual and the organisation shall be identified and analysed.

2.   Targets and standards shall be set for the trainee, which are within his capabilities.

3.   The pace of the training programme should reflect the trainee's ability to maintain progress in properly absorbing the same.

4.   The trainee shall receive regular feedback of results. Any problem areas shall be highlighted, discussed and resolved.

5.   As the trainee progresses the amount of information provided shall be gradually reduced, thus inducing a feeling of independence and competence.

It is common place to find a wide variety of tasks in an organisation and each will require varying degrees of skill, effort and responsibility. This being so, it is inevitable that rates of pay will also vary and the differentials between the jobs will reflect their relative values. However, other factors such as local market conditions, bargaining strengths and traditions also influence a company's payment structure and a great deal of planning is required if rationalisation is to be achieved. One technique which has been successfully adopted by many companies to establish an equitable wage structure is job evaluation. In a job evaluation exercise a comparison is made of common criteria over a range of jobs, and the resulting analysis may be linked to a points allocation or job ranking system, and hence to a wage scale.

In conducting a job evaluation exercise it is important to cover a reasonable variety of tasks within the whole spectrum. For each, a job description is prepared setting out details of the duties and responsibilities undertaken by the employee together with a statement about his working conditions. Very often this task is undertaken by work study personnel since, they are responsible for determining methods of operation and evaluating the work content of the job. Each job is assessed factor by factor, resulting in a comprehensive comparative analysis.

The individual is the most important resources of any company and only people who are well trained, well motivated and adequately rewarded will provide a positive and synergistic contribution towards the company's objective and its organisational health.

In most cases, the factors, which may be weighted according to relative value, are as follows:

1.   Skill-education, experience and training.
2.   Effort-both physical and mental.
3.   Responsibility-for equipment, materials, initiative etc.
4.   Working conditions-general conditions, risk of accident and injury.

Pay policies affect not only individual employees but the whole organisation, and the rewards and objectives vary at different levels within the enterprise.

Industrial relations

An effective industrial relations policy is important, since is the system through which employees take part in decision-making and in many instances it affects the while atmosphere of employer/employee relationships. An industrial relations policy is essentially a set of rules whose determine procedures for negotiation on such matters as:

1.   Wage and salary scales.
2.   Terms and conditions of work.
3.   Disputes and grievances.
4.   Recruitment and dismissal.

5.   Other issues of mutual interest, e.g. closed shop, redundancies and joint consultation.

In order to promote an atmosphere of co-operation, and to minimise conflict, the needs of management and work people must be recognised by both sides. Trade unions exit to protect the interests of their members and improve their working conditions. Management, while aware of the pressures and constraints imposed by the trade unions, have a duty to maximise the use of resources at their disposal, which may be expressed in relation to profitability, return on investment, level of service, sales volume, market share and cost-effectiveness. The strategies adopted in attempting to solve industrial relations problems will vary from company to company, and indeed from union to union, but there is no doubt that they will be influenced by both internal and external factors.

Internal factors

1.   The attitudes of employees to management, and management to employees.
2.   The leadership style of management.
3.   The bargaining strength of both parties.
4.   The number of negotiating bodies.
5.   The prosperity of the company.

External factors

1.   The extent to which parent boards influence company management, and district officials influence or control local shop stewards.

2.   Whether or not bargaining is conducted at plant, local or national level.

3.   Government policy towards industrial relations.

4.   The economic situations nationally, locally or within the company itself.

3.5   Describe the project implementation schedule.

Implementation of this project is a challenging task and calls for meticulous planning, scheduling and monitoring to realise the project goals within the budgeted cost and time frame. The goal can be achieved by adopting modern project management techniques.

To implement this project adequately, a team of engineers and project personnel having requisite education and experience are being appointed, to whom a detailed Work Breakdown Structure (WBS) in a logical order of activities, shall be supplied shortly, keeping in view cost estimation, scheduling, and to help monitor and control of the project. It is proposed to be formulated in conjunction with the objectives of each activity and goal settings. Project shall be programmed and controlled by network analysis techniques. Before the application of the network analysis techniques, the project personnel shall be acquainted with their capabilities in saving time, resources and costs. The training of project personnel and engineers at levels shall be provided for proper control of project progress and taking of timely corrective actions to re-align these efforts to meet ore stated objectives. It is proposed to gear programming and control system, i.e. project implementation system, which will ensure an integrated approach to project implementation. Project management activities shall be determined in advance and all activities carried out be project personnel as well as those to be contracted shall be identified.

Responsibility for project implementation shall be clearly defined. The forms of project organisation range from project oriented to functional organisation, while most of the cases are combinations of the two, with certain adaptations to prevailing conditions. It is impossible to over emphasise the importance of establishing a team of a task force for implementing the project with a designated leader to co-ordinate and guide its functions.

Project manager: Shall be responsible to the Board of Directors. The project manager shall be responsible for guiding and co-ordinating the efforts of all parties engaged in implementing the project, obtaining necessary government approvals on contracts. He is to control the project organisation with the promoters as well as with other agencies and organisations interested in the project. The manager shall have some staff to assist him, especially in checking expenditures to date and determining the present and future cost overrun or under run so that the project manager can take or propose to the Board pertinent corrective measures.

The network shall cover the pre-construction phase of the project indicating major administrative processes, since experience shows that some of them have frequently involved lengthy delays. In other words, it shall include the aggregate activities to be carried out be the principal parties participating in the implementation process.

The project budget shall then be prepared. Part or periodic payments to contractors which might be made at the end of certain intervals (e.g. weekly or monthly) throughout the time horizon of project implementation, shall be made by summing up activity costs per unit of time, which may be a week & month, and computing the cumulative cost at the end of each time interval. For the activities that are in process and are contacted or sub-contracted, the assumption of a linear time cost activity relationship shall be used for the sake of simplicity. In other words, expenditures are uniformly distributed throughout the duration of the activity.

PROJECT SCHEDULE

After an investment decision is taken, the main machinery and long delivery items must be ordered out at the earliest, forming the first major step in implementation of the project. It is foreseen that an engineering consultant will be appointed for carrying out the detailed engineering including basic engineering and procurement assistance to the client. It is also assumed that reputed and experienced contractors with adequate resources viz., men, materials, tools and tackles etc. will be engaged for execution of the construction and erection work. The purchase packages for auxiliaries shall be kept minimum so as to reduce the co-ordination efforts to the minimum. A great deal of co-ordination is required for constructing/erecting the new units. This task is feasible, provided the major activities of the project are co-ordinated and completed in the duration specified to achieve the respective milestones in time.

STRATEGY FOR TIMELY EXECUTION

It is important to deploy a team of experienced personnel for project execution and select the external agencies with due care for rendering the services and supply of equipment for the project. The project activities must be identified, planned and scheduled, and the progress monitored for timely project implementation. All the inputs to the project including financial resources must be identified and their inflow planned and arranged in time.

The project must be managed professionally with necessary co-ordination among the various agencies and requisite decisions taken promptly.

Establishment of an effective monitoring procedure for progress review and co-ordination.

In short, the following key factors would constitute the broad strategy for timely execution of all activities in a pre-determined manner as per schedule shown in the bar chart as to reach at a basis of regular production.

Early selection of an effective in-house technical team (TASK FORCE) by Government of Timor Lorosae, headed by a Project Manager for planning and executing the project.

i)   Proper choice of external agencies such as consultants for Project Engineering., Machinery suppliers, Construction Agencies etc. keeping in view their reputation/past performance and working experience in their fields.

ii)   Adequate use of computer-based PERT/ CPM techniques for project planning, scheduling and monitoring.

3.6   Description of the initial project Flowchart.

The project implementation phase embraces the period from the decision to start the project to the beginning of the commercial production. It includes a number of stages including negotiations and con-tracting, project design, construction and start-up.

3.7   Description of the requirement of Land.

For this project the Government of Timor Lorosae will participate with about 25.000 square meters of land. In selecting the land the following criteria should be followed.

1.   The land should be near to the main city of Dili.
2.   The land should be properly connected by road.
3.   It should also have proximity to many end user clients.
4.   It should also meet various Government policy of
        achieving the social objectives.

2.08 Description of the requirement of Building.

The total requirement of building is as follows.
a. Adm. Building   400 Square Meters
b. Studios −2 (Two)   1000 Square Meters
c. Other Misc. Building   200 Square Meters

Total Square Meters: approximately 1600 Square Meters.
Cost for construction of about US$.750 per square meter (including air conditioning systems, primary electrical cabin etc.).

2.9   Description of some requirements for Plant & Machinery and their basis of selection.

The plant and machinery have been selected giving due consideration to the sophisticated nature of technology required. Detailed discussions were carried out with the foreign suppliers to ensure that the required capacities are practical with minimum capital and operating costs before the machinery were finally selected.

The machinery shall be selected from the most reputed foreign manufacturers of complete range of equipment. Keeping in view the following main factors:

a)   Past performance
b)   Existing machinery in Africa & abroad.
c)   Sales and service facilities in Africa

The main equipment is as follows.

STUDIO

1.   Post-production equipment – non-linear post-production, two ES7 stations.
2.   ENG – 2 DVCAM Camcorders
3.   Régie with 2 DXC-D30PK1 Digital Cameras - for cold light installation.
4.   Flexicart.
5.   Computer graphics suite.

Other technical data:
The whole installation is characterised by:
1. Informational web with UTP level, 5 cables.
2.   Phone system.
3.   Specific furniture.

Discrimination of Technical Data:

Studio

Cold lighting system.
2 Camera channels composed by:
1   DXC-D30PK Sony Digital Cameras 2 J18 Canon Zoom Lens
2   CA-537P Sony Camera Adapters
1   DXF-50 Sony Studio Viewfinder's
1   CCU-M3P Sony Control Camera Units 2 RM-7P Sony remote CCU controls
2 50M Sony Multicore Camera Sets
2 Canon Focus and Zoom control sets
2 Intercommunication Headphones
2 Special tripods
1   charriot plane and curve
1   Teleprompter with 2 reading systems (monitors for individual speakers) Different microphones with diverse directional characteristics

Audio components Video components

Video Régie
1 Recorder/Player BETACAM equipment
1 video digital mixer with 12 tracks
1 DVE equipment of digital effects for 3 channels
1 TBC Frame Synchroniser
1 Oscilloscope and Vectorscopy for control of cameras
1 Synchroniser Generators with Changeover
1 Intercommunication system with 4 places
1 Audio and Video Matrix with remote controls Video distributors

Audio Régie

1 audio mixer with 24 tracks
1 audio monitors
1 cd player
1 cassette deck
1   DAT
1   Mini-Disc
1   Level Detector for Audio Stereo Stereo Audio Distributors

Audio compressors Audio effects Microphones

Video Post-Production
2 Sony ES-7 on DVCAM/AVID hybrid non linear edition stations 1 Conventional BETACAM Edition Suite 2:1
Multi-track audio (8) on hard disc, Sound Scape type

3 ENG SETS:
Sony DSR-300 PK DVCAM Camcorder Digital Compact Report projectors
Various support materials (Batteries etc.)

3.10   Description of the Water Requirements for the project.

The requirements of water, separately for various matters are given in the following table.

Circulating   : Nil
Make-up   : Nil
Process   : Nil
Drinking   : 3,000 LPD

3.11 Description of the Steam requirement for the project.

A. Steam requirements and steam balance   : N.A.

B. Capacity and type of boiler with detailed specifications   : N.A.

C. Steam and energy diagram : N.A.

D.   Total energy generated / purchased (converted into M. K. : N.A. Cal) theoretical requirement of energy (in M. K. Cal)

at the various consumption stations and expected actual requirement at these stations.

E. If alternate processed are available, comparative energy : N.A.

consumption figures for the various processes. If the
project is energy intensive, possibility of choosing alternate process in order to make the project less energy
intensive.

F. Steps proposed to be taken by the company to improve   : N.A.
energy losses efficiency and reduce energy losses (such
as power factor improvement, power load management,
optimising, illumination waste heat utilisation, etc.)

G. Scope for usage of solar / other renewable sources of   : N.A.
energy.

H. Any other measures contemplated in the direction of : N.A.
energy conservation and management.

3.123 Some information on Compressed air, fuel, etc.

Compressed Air

(a) Requirement   : N.A.
(b) Sources   : N.A.

(c) Arrangements proposed   : N.A.
(d) Cost at site with detailed calculations : N.A.

3.13   Details of the nature of atmospheric, soil and water pollution likely to be created by the project and the measures proposed for control of pollution. Indicate whether necessary permissions for the disposal of effluent have been obtained.

There shall not be any atmospheric pollution likely to be created by the project as there is no machine which has combustion resulting into air pollution or any chemical process which may release any gases which may result into an air pollution.

However, the use of DG set shall cause a small amount of air pollution. Considering the size of DG set, the pollution is within the permissible limits.

TIMOR LOROSAE TELEVISION NETWORK

THE INDUSTRY

CHAPTER - 4

THE INDUSTRY

4.01 Description of the Television Industry in general.

The television industry can be described in the following broad categories.

1.   Introduction

2.   Cable Networks

3.   DTH

4.   TV-media

5.   Earnings drivers

6.   Outlook

The detailed discussion in each of the categories is given below.

Introduction

The growing popularity of TV as a communication medium has resulted in the TV media sector under-going a rapid transformation. From the black and white days of state controlled TV Station, to the highly colourful tunes of Channel V and MTV, the medium has certainly undergone a phenomenal change. Given its popularity, percentage ad spend has also increased proportionately on this medium.

Media pie (%)

	1995	1997
TV	62.5	68.8
Radio	20.9	15
Press	16.6	16.2

Source NRS

Entry of new channels post 1991

All over the world the telecasting has witnessed entry of new channels to cater to the various needs of world audiences. Channels have been launched in English as well as other regional languages. In many countries of the world till 1991, the state owned TV Station ruled the roost, as other players were not allowed to uplink and broadcast. However channels such as CNN, Star TV and BBC, which were offshore companies, could circumvent these regulations and telecast their programs into any country of the world. Cable operators then relayed the same and made it available to the common man through the cable television network.

Like many other countries, the State machinery controlled television. It was used as a propaganda tool for the party in power, with the opposition always at the receiving end. The customer had very little choice. The first steps towards more user choice began during the 1980s, which had to be telecast to a wider audience. TV Station used satellite channels for the telecast and the T.V. network was launched as an international channel.

The sports telecast by Channel 9 in 1985 and the

Gulf War in the late eighties all played small but important cameos in educating the international viewer. With liberalisation in 1992 and crumbling tariff barriers televisions (read as colour TVs) became more easily available. The media revolution had started.

Major satellite channels avidly watched by viewers are Star TV, Sony TV, Home TV, BBC & CNN. There are other regional language channels which are major players in their respective territories.

Most of the channels that could not attain popularity rapidly suffered, since their advertisement earnings were not sustainable. The first round of the media wars is over. Management changes, i.e. original promoters selling out to new management with deeper pockets, has become the order of the day. Alliances like the famous ESPN Star Sports arrangement also made headlines. Given the global trends of mergers and acquisitions, further consolidation is likely. Alliances and mergers make sense when the partners complement each other, like BBC and Discovery launched Animal Planet, CNBC and ABNI came together to launch a business channel called CNBC Asia.

Cable Networks

Antennas set up by either the end user or the cable operator receives the signals transmitted by the satellite. Local cable operators lay their own cables, set up control rooms, which can telecast 40 or more channels over a limited area. They charge the household a one- time connection charge of about US$.10 per point and a recurring monthly charge ranging between US$.1 to US$.5.

Initially, this was done in a very unorganised manner. The business required local knowledge and contacts, so every locality had its own cable operator. Collection was critical for the cable operator. For the end user, quality of telecast and a complete lack of standards became an issue. This lead to a shakeout and the formation of cable companies with money power which in turn tied-up with the local and small cable operators. Cable companies charge about US$1 per month to the local cable operators and support them with training and other infrastructure inputs. The business is immensely capital intensive and takes a long time to break even.

In many countries the operations of cable operators are regulated under the Cable TV Act which ensures that pornographic materials and other materials which are against culture and values or are detrimental to national interests do not get telecast. Recently this act has been amended to include foreign channels also.

Direct To Home

DTH is a new technology that circumvents the cable operators by directly delivering a bundle of channels to the end user. DTH involves transmission of encoded audio/ video signals (Ku band) via satellite. The end user needs an antenna to receive the signals and a decoder (set top box) to unscramble the encrypted signals. DTH services elsewhere in the world are Echostar and DirecTV (USA) and BskyB (Europe). Rupert Murdoch of Star TV fame owns BskyB.

The size of the antenna in DTH will be 1.5-2 ft in diameter, making it easy to install and transport. In conventional cable, since signals are in C band, an 8ft- diameter antenna is needed. The basic difference in the business model is the hardware costs in DTH. In a cable system, the user pays a one time connect fee and monthly rentals, while in DTH he has to invest in hardware.

The antenna will cost about US$200-300 and decoder will cost about US$200. The African viewer might be reluctant to incur such heavy installation costs. Quality of telecast in DTH is superior to Cable TV and viewer can receive up to 200 channels.

DTH will result in restructuring of the cable television industry. It will become imperative to have cash reserves to withstand the technology threat. Up gradation to fibre optic backbone will become necessary. A fibre optic network will cost about US$0.5mn per km as compared to US$0.1mn per km for coaxial cable. The stage is now ripe for consolidation.

TV Media features

In the Broadcasting business, it is only the industry leader who makes sizeable profits. The business is a game of asymmetrical payoffs. For instance, the top 5 channels account for 90% of ad spend.

Urbanisation and TV penetration is related. This may be due to the popularity of cable television that has resulted in increased colour TV sales. Rural penetration is low, although growing at a fast pace, because of dearth of specific program content to cater to that segment.

Liberalisation has resulted in the world viewer becoming more aware and conscious. This has resulted in the customer having more choice with the entry of a number of companies in different segments. Competition has resulted in companies increasing their marketing spend significantly.

Popularity of TV media is becoming higher. Increasing TV penetration leads to a reallocation of advertisement budgets with higher allocation for television at the cost of other medium.

TV channel operators use different business models to generate revenues. The critical component of any channel is the quality and type of programs they telecast. This determines their popularity, which in turn determines amount of advertisement revenues they can generate. They can do any one of the following:

Buy programming rights of program software from outside and collect advertisement revenue on their own. This model is followed by several TV companies, wherein they have a separate company in their fold, which develops all the content. The advantage is that re runs of serials/ programs become very profitable.

Selling time space to the producers for a fixed charge. Producers in turn are free to book advertisements at their own rates (there is an understanding on the time allocated for advertisement) and collect revenue. This is the basic model for many TV companies in which they sell prime time slots. The rights continue to be vested with the producer.

Earnings drives

The key factors that drive sector revenues are

Television penetration: Since the medium is television, increased television penetration will imply higher viewer ship. This will translate into higher advertisement spend allocation. This will also imply higher software production and demand for new programs.

Competition from other satellite channels would have an adverse impact on advertisement revenues, as advertisers have more choice in allocating ad budgets.

Government policies can have a big impact on the fortunes of the entire industry. When the DTH bill is passed in any country then, it will trigger a restructuring of the cable business.

Launching new channels targeted at specific segments, like regional channels within any country other areas having large pockets of ethnic population would lead to revenue growth. This will entail significant initial outlays.

Depreciation of the local currency would increase revenues as most of the program/ software companies export the programs overseas and payments are dollar denominated.

Advertisement revenue

As mentioned earlier, this is the primary source of

income for TV channel operators. This revenue is directly co-related with the reach and viewer ship of a channel. Any channel's popularity depends on good quality programs, which is the software content. The business requires enormous initial investment in programs and revenues follow only with a time lag after the channel receives a minimum viewer acceptance.

Outlook

The sector has latent potential for growth on back of the exponential growth of cable TVs during the last 5 years. Television penetration in Africa is extremely low as compared to other developing countries like Malaysia, Pakistan, etc. in Asia. The number of channels has increased, implying higher demand for software programs.

Advertisement revenues, which are the barometer of channel popularity, will get dispersed over several competing channels. A shakeout is likely in both the channel and cable TV sectors. The biggest beneficiaries will be the content providers or the software houses. They will control the intellectual rights to the key element driving any channel's popularity.

Direct -to-Home, Digital Terrestrial Transmission and Conditional Access Cable Delivery have emerged as new delivery mechanisms. Breakthrough in technology would help open up avenues for these channels.

TIMOR LOROSAE TELEVISION NETWORK

MARKETING

CHAPTER - 7

MARKETING

7.1   Description of the commercial viability of the project with regards to revenue generations.

Estimations shows that, after about 5 years of regular working, the CPLP Cultural Television Channel will be able to start exporting a great quantity of services and videos to Africa, Europe and South

America. (See financial projections)

The economic feasibility of the project will occur with the commercialisation of several products, like:

1.   The sale and production of cultural programmes to various institutions like:
-       BBC
-       TV Culture Brazil
-       RTP
-       Discovery
Etc.
All over the world commercial networks for television cultural programmes have been created. Several television channels in different countries have regularly acquired programmes by producers spread out all over the world. Not only, several international organisations, like the World Bank and the United Nations, including FAO and Unesco, among many others, need programmes for public education, like programmes oriented to alert and to educate people concerning endemic and epidemic diseases. AIDS, malaria, tuberculosis are a very few examples.

These world institutions have made a great effort to develop television programmes with humanitarian objectives. But, in Africa – undoubtedly the continent most needed of such programmes – there is no television station, in present times, with capability to make face to such a need.

Therefore, the television teams that had been responsible for such programmes are, practically in its totality, placed in countries of the called First World, strange to the local population's concrete reality.

Programmes focusing new agricultural techniques or even oriented to agricultural, commercial and industrial education are essential elements of such a repressed demand.

The price for each institutional campaign, with an average of 10 films produced per campaign, is of about USD.15.000$ and the capacity of the CPLP Cultural Channel will be as follows:

First Group* - from 2 to 4 campaigns per month.
Second Group – from 1 campaign in two months to 1 campaign per month

This represents:
First Group   a potential annual income of about US$.540.000
Second Group a potential annual income of about US$.90.000

*First Group: Angola, Cape Verde, and Mozambique. Second Group: Guinea-Bissau, Sao Tome e Principe, East Timor (Lorosae). Reference Group:

Brazil, Portugal.

Obs. For the case of Guinea-Bissau and East Timor, it is expected a decrease of such incomes for US$.216.000 and US$.36.000.

2. The sale and production of no cultural programmes to other television channels, like:
-        series (novels)
-        talk-shows Etc.

Many countries of the region need to produce television programmes, but do not have capability to do it. Thus, they seem themselves obliged to search expensive productions in Europe. With only one programme sold-three daily hours-the channels of the First Group* would receive in incomes the equivalent of about two times of the whole investment. The Second Group countries will not have conditions at the beginning to product cultural programmes to other television channels.

*First Group: Angola, Cape Verde, and Mozambique. Second Grope: Guinea-Bissau, Sao Tome e Principe, East Timor (Lorosae). Reference Group: Brazil, Portugal.

The price for this type of television programme is:

Auditory programs   US$.12.500 per programme
Talk-shows   US$.4.000 per programme

Interviews   US$.4.000 per programme

The capacity of production, of these programmes by the countries of the First Group will be:

In the FIRST PHASE:
Talk-shows   2 per month
Interviews   4 per month
In the FINAL PHASE:
Auditory programmes   6 per month
Talk-shows   20 per month
Interviews   20 per month

Thus, the potential income will be:

FIRST PHASE
Talk-shows   US$.96.000 per year
Interviews   US$.192.000 per year

FINAL PHASE
Auditory programmes   US$.900.000 per year
Talk-shows   US$.960.000 per year
Interviews   US$.960.000 per year

Obs. All values must be decreased for Guinea-Bissau and East Timor in the first years, as it is referred in the detailed feasibility studies.

The price of the novels is much higher. For each complete novel, with about 120 chapters, the price is about US$.2.500.000 and the capacity of the First

Group of the CPLP Cultural Channel (in the final phase) will be of one novel per year.

3. There is the possibility of commercial use of the transmission time beyond the six hours reserved to culture and education.

The conventional day in television is of 18 hours, of which only 6 hours would be no commercial. We would have, therefore, 12 hours of no cultural transmission, which should be freely commercialised.

Even the period of six hours- divided into two sections of three hours, the first one dedicated to culture and second section of three hours to education-has a great potential for sponsoring.

Each commercial hour of transmission can include up to 12 minutes of advertising. The price for each 30 seconds of advertising is:

Noble time   from 6PM to 11PM   a b o u t US$.400 each 30"
Normal time   rest   about US$.200 each 30"

Concentrating the commercial programming in the noble time. With 4 hours of transmission in this period, we would have:

Educational television: from 7AM to 9AM

From 3PM to 4PM
Cultural television: from 4PM to 7PM
Commercial television: from 9AM to 3PM
From 7PM to 1AM

Thus, the commercial period would comprehend 6 hours in normal time, 4 hours in noble time and 2 hours in normal time again.

Begin 12 minutes per hour for advertising, the commercial period could have:

Noble time   48 minutes   24   films   US$.9.600 per day
Normal time   96 minutes   48   films   US$.9.600 per day
Total of the potential income............ US$.19.200 per day
                    or US$.576.000 per month
                    or US$.6.912.000 per year

It is believed that in the first phase of the project (after one year), the CPLP Cultural Television Channels of the First Group will be able to start with an income from commercial advertisement of about US$.1.000.000 per year-value which is predicted to increase in the follow months.

4. The rent of the studios to television teams of other countries can be another source of incomes.

Many producers who cover events in Africa need to move many times to their countries of origin during the video works, because there is not technical support in Africa. The same phenomenon happens with the cinematography and the journalism productions. It is not difficult to imagine, for example, the serious problems journalism teams have had, for example, with essential components like batteries, lighting, electronic components etc., which only can be easily find in Europe.

The period of renting of a television studio is of 12 hours. Each period has price of about US.4.000. After the first year, the studio should be rented for 12 hours per each period of 3 days. So, the rent per month, in this period (First Phase), will be able to generate incomes of about US$.120.000 per month in the First Group.

5. Support services to other television networks in all areas, including novels, mini series, docdramas series etc.

Many countries, principally in Africa, do not have technical conditions to develop this kind of programmes, but they have a strong internal repressed demand in this sense.

Each team, abroad, has a price of US$.1.000 per day (two people), after the costs of dislocation, hotel and meals. In the first months (first phase) it will be organised one team for works in other countries. In the final phase it is predicted to have up to 5 teams with such function. The capacity of each team is of about 20 days per month. Therefore, the support to other television networks-when took in its full potentiality-will be able to represent up to 20 days per month in the first phase. This represents a potential income of about US$.240.000 per year, and 100 days per month of works in the final phase, signifying about US$.1.200.000 per year of incomes, always referring to the First Group countries.

6. Colloquies, seminars, videoconferences and meetings of different natures.

The building of the CPLP Cultural Television Network will have a medium size auditorium, (First Group), with all conditions to receive seminars, colloquies and meetings of the most diverse nature. Such seminars, videoconferences etc. are important not only for the increase of the incomes, but also to attract specialists of the most different areas, being an important element for the development of all region as well as for the diffusion of the local, regional and continental cultural values.

The meetings, seminars and colloquies attract, in average, about 300 people per event. The price-excepting meals, hotels and transportation-per each participant is of about US$.50 per day. It will be a capacity for up to four events of this type per month, what could represent an income of about US$.60.000 per month or US$.720.000 per year (final phase-after 5 to 7 years).

The project should also turn possible.
-        classes of professional formation in diverse disciplines
-        support for a multimedia high technology centre (First Group)
-        support for a multimedia high cultural centre (Final Group)

TIMOR LOROSAE TELEVISION NETWORK

PROFITABILITY & CASH FLOW

CHAPTER - 8

PROFITABILITY AND CASH FLOW

8.1   Estimation of cost of production and working results for the first five years of operation.

The estimated cost of production of working results for the first five years of operation are given in the chapter "Financial Projection".

8.2   Cash flow statement for the company as a whole, for five operating years of the project based on the estimates of working results.

A detailed cash flow statement for the company as a whole for five operating years is given in the chapter of "Financial Projection".

8.3   Projected balance sheet for five operating years for the company as a whole.

The balance sheet for five operation years for the company as a whole is given in the chapter of "Financial Projection".

TIMOR LOROSAE TELEVISION NETWORK

ASSUMPTIONS

CHAPTER - 9

ASSUMPTIONS

1.   The Timor Lorosae Television Network shall generate the income from the following sources.

a)   Sale of cultural programmes.
b)   Sale of commercial programmes.
c)   Sale of advertisement time during transmission of TV programmes

d)   Hiring of studios.
e)   Supply of technical services
f)   Hiring of conference hall

The detailed assumptions for each of the above mentioned activities are as follows:

a)   Sale of cultural programs:

I.   It is assumed that four cultural programs shall be produced per month in the first year, which can be sold to other countries. This

figure will increase to 5 programs per month in the second year, six programs per month in the third year and so on.

II.   It is assumed that the selling price shall be US$.60 per program.

b)   Sale of commercial programs:

I.   The commercial programs consist of talk shows, interviews and auditory programs.

II.   It is assumed that two talk shows shall be produced per month in the first year, three talk shows shall be produced in the third year, four talk shows per month shall be produced in the third year and so on.

III.   It is assumed that the selling price of one talk show program shall be US$.1600.

IV.   It is assumed that four interviews shall be produced per month in the first year, five interviews shall be produced in the third year, six interviews per month shall be produced in the third year and so on.

V.   It is assumed that the selling price of one interviews program shall be US$.1600.

VI.   It is assumed that no auditory programs

shall be produced in the first and the second year. Only when the people have two years of experience then in the third year one auditory program shall be produced per month. In the fourth year two programs shall be produced per month and so on.

VII.   It is assumed that the selling price of one auditory program shall be UD$.5.000.

c) Sale of advertisement time during the transmission of T.V. programs:

I.   There shall be a total transmission of 8 hours per day in the first year. It will go on increasing by two hours in the second and subsequent years.

II.   Out of which advertisement shall be available for programs with transmission period of four hours in the first year. The programs in

which advertisements could be available shall increase by two hours per year in the second and subsequent years.

III.   In one hour T.V. transmission, 12 minutes of advertisement shall be allowed.

IV.   Out of total TV time 33% time shall be considered as prime time and remaining 67% shall be considered as non prime time, in the first, second and third years. From the fourth year onwards the prime-time advertisements shall remain constant.

V.   The advertisement rates shall be US$ 320 for one minute of prime time advertisement and

US$ 140 for one minute of non prime time adver-
tisement.

d)   Hiring of studios:

I.   It is assumed that the studios shall be taken
on hire for a shift of 12 hours, two times in a month
in the first year. It will increase to three times in a
month in the second year.
II.   The rent per day shall be US$.1.600.

e)   Supply of technical services:-

I.   It is assumed that one team, consisting of
two technically qualified people shall be available
in the first year. In the second year two such teams
shall be available. In the third year three such teams
shall be available and so on.
II.   It is assumed that in the first year the team
shall be hired for 10 days in a month. In the second year the teams shall be hired for 12 days in a
month. In the third year the teams shall be hired
for 14 days in a month and so on.
III.   The rate of hire shall be US$.400 per team
per day.

f)   Hiring of conference hall :-

I.   It is assumed that the conference hall shall
be taken on hire for two times in a month in the first
year, three times in a month in the second year and

four times in a month in the third year and so on.

II.   The hire charges shall be US$.200 per conference.

2.   The unit is planning transmission of three hours of educational programs, three hours of cultural programs and two hours of commercial programs per day in the first year. Thus, there shall be a total transmission of eight hours of programs per day in the first year. The total transmission hours shall increase by two hours per day in the second and subsequent years. Thus, in the second year there shall be total transmission of 10 hours per day and in the third year there shall be a total transmission of 12 hours per day. The cost of software production is assumed at US$.1.000 per hour.

3.   The cost of miscellaneous consumable items is assumed at US$.14.400 in the first year. This will increase by 10% in the second and subsequent years.

4.   The cost of stores and spares consumed is assumed at US$.16.800 in the first year. It will increase by 10% in the second and subsequent years.

5.   The cost of repairs and maintenance is assumed at 0.1% of the total cost of plant and machinery in the first year. This cost will increase up to 0,3% in the fifth year.

6.   It is assumed that the unit shall use 1.000 units of power every day or 365.000 units of power in the first year. The consumption of power shall increase by 10% in the second and subsequent years. The cost of power is assumed at 2 cents for one unit.

7.   It is assumed that the unit shall consume 1.000 litters of diesel per month. The cost is assumed at 13 cents for one litter. It will increase by 10% in second and subsequent years.

8.   The unit shall have a technical staff of 35 people and administrative staff of 15 people.

9.   Depreciation is calculated on reducing balance method. The rate of depreciation is taken at 10% for building, 25% for plant and machinery, 10% for furniture and fixtures and 20% on vehicles.

10.   Advertisement expenditure is assumed at 0.2% of the sales.

11.   It is assumed that the stock of raw material and stores shall be 30 days of consumption, receivables shall be 45 days sales and creditors shall be 15 days purchases.

12.   The term loan shall be for a total period of 25 years. It shall have a moratorium of five years. It shall be repayable in 20 equal yearly instalments.

brief bio

Emanuel Dimas de Melo Pimenta (1957) has been considered by many as one of the most interesting architect, musician, photographer and intermedia artist at the beginning of the third millennium – according to statements written by like John Cage, Ornette Coleman, Lucrezia De Domizio, Merce Cunningham, John Archibald Wheeler, René Berger, Dove Bradshaw, Daniel Charles, Phill Niblock or William Anastasi among others.

His works are included in some of the most important art collections and world-wide recognized institutions like the Whitney Museum of New York, the ARS AEVI Contemporary Art Museum in Sarajevo, the Biennale of Venice, the Cyber Art Museum of Seattle, the Kunsthaus of Zurich, the Durini Contemporary Art Collection, the Bibliotèque Nationale of Paris and the MART - Modern Art Museum of Rovereto and Trento among others.

He has developed architecture, urban projects and music using Virtual Reality and cyberspace technologies and neurosciences.

His works are included in the *Universalis Encyclopaedia* (Britannica Encyclopedia) since 1991;

in the *Sloninsky Baker's Music Dictionary* (Berkeley); the *Charles Hall's Chronology of the Western Classical Music*; in the *All Music Guide – The Expert's Guide to the Best Cds;* the *Wikipedia* and in the *Babylon* among others. Articles on his works have regularly appeared in different newspapers and magazines, like *The Wire, Ear, New York Times, Le Monde, Le Parisien, Liberation, O Estado de Sao Paulo, O Expresso,* and *O Globo, Il Sole 24 Ore* and *la Reppublica*, among others.

In the early 1980s Emanuel Pimenta coined the concept "virtual architecture", later largely used as specific discipline in universities all over the world. Among his challenging architectural designs, there is the famous floating island for Lisbon, Portugal; an experimental floating island for the Lake Maggiore in Switzerland; the Symmetrion building in Budapest, Hungary; and the Time Design Museum building, in Trancoso, Portugal.

In the end of 1980s he published in England the first book of the world on virtual architecture.

In 1980 he started elaborating *Woiksed* - the first virtual planet of the world, anticipating *Second Life* in about twenty years. In 1994 he received an important European Prize for that project.

In the 1990s he was the curator of the first exhibitions of virtual architecture all over the world. One of these exhibitions was at the Biennale of São Paulo, in 1999 and 2000.

In the year 2000 Emanuel Pimenta starts his works and researches on space architecture. Among his designs in this field we can underline Kairos, an orbital building, and UIRA – the first orbital Olympic Village in history.

As a composer, since the end of the 1970s, he has developed graphical musical notations inside virtual environments.

His concerts integrating visual art have been performed in various countries in the last thirty years. One of the most important moments was his famous concert at the Biennale of Sao Paulo, for four large orchestras, in 1985, side by side with John Cage, Robert Rauschenberg, Nan June Paik and Bill Viola among others. In 2008 he created the first opera on Dante Alighieri's *Divine Comedy* in the history of music, with the world première at the Abstracta Festival, in Rome, Italy. In 2009, his concert *CANTO6409*, created in partnership with the Italian movie director Dino Viani, who was responsible for the film, has its world première at the International Film festival of Cannes, in France.

Legendary musicians like John Cage, David Tudor,

Takehisa Kosugi, John Tilbury, Christian Wolff, Martha Mooke, John DS Adams, Maurizio Barbetti, Michael Pugliese, Umberto Petrin, Susie Georgetis, Audrey Riley and the Manhattan Quartet among others have performed his compositions.

He collaborated with John Cage, as commissioned composer for Merce Cunningham, from 1985 until his death in 1992. He remained composer for Merce Cunningham in New York City until his disappearance in 2009.

His concerts have been performed in some of the most prestigious theatres all over the world, like the *Lincoln Center* and *The Kitchen* in New York; the *Opera Garnier* or the *Theatre de La Ville* in Paris; the *Shinjuku Bunka Center* in Tokyo, the Montpellier Municipal Theatre, the Festival of Aix en Provence, the Modern Art Museum MASP in Sao Paulo, *La Fenice* in Venice, and the Biennales of Venice and São Paulo among others.

Since 1989, he has regularly collaborated with Lucrezia De Domizio, the Baroness Durini, with countless contemporary art projects, concerts, exhibitions and publications – especially on Joseph Beuys – in several countries.

In 1987 he started a regular collaboration with the legendary Swiss art philosopher René Berger in numerous projects especially in Switzerland – with Nan June Paik, Bill Viola, Edgar Morin, Pierre Levy and Basarab Nicolescu – until his death in 2008.

In the 1990s, Pimenta develops a gigantic project for a planetary television network oriented to culture and art.

With more than four hundred musical compositions already recorded, more than forty published compact discs, four cd-roms, he has wrote and published more than seventy books, great part of them individually, several papers and a great quantity of electronic books. His works have been regularly published in England, the United States, Japan, the Netherlands, Portugal, Brazil, Germany, Canada, Switzerland, Hungary, Italy and Spain.

He has also been curator for various institutions, like the Biennale of Sao Paulo, in Brazil; the Calouste Gulbenkian Foundation and the Belem Cultural Center, in Lisbon, among others.

He won the National Marketing Prize in 1977 by the Brazilian Association of Marketing; the APCA Prize in 1986 by the Art Critics Association of Sao Paulo (AICA Section in São Paulo); and the Lac Maggiore Prize in 1994 by UNESCO, AICA International Association of Art Critics, the Council of Europe and the Lombardia Regional Government. In 1993 his works were selected by UNESCO, in Paris, as one of the most representative intermedia researchers of the world.

He is member of the *SACD – Societè des Autheurs et Compositeurs Dramatiques* in Paris since 1991. He also is an active member of the *European Environmental Tribunal*, in London, where he has been member of the board since 1995.

He is an active member of the *New York Academy of Sciences*; of the *American Association for the Advancement of Science* in Washington DC; of the *ASMP - American Society of Media Photographers*; of the *CAU Brazilian Council of Architecture and Urban Planning*; and of the *OA Portuguese Order of Architects* among others. He is member of the Space Architecture Technical Committee of the American Institute of Astronautics and Aeronautics. He is member and advisor of the *AIVAC – Association Internationale pour la Video dans les Arts et la Culture*, in Locarno, Switzerland. He was a founding member of the *International Society for the Interdisciplinary Study of Symmetry – ISIS Symmetry* and of ISA *International Symmetry Association*, both in Budapest.

He was Editorial Director of the art and culture magazine *RISK Arte Oggi*, in Milan, founded and directed by Lucrezia De Domizio, Baroness Durini, from 1995 to 2005. He was also member of the Advisory Editorial Board of the art and science magazine *Forma*, in Tokyo, Japan; and he is member of the Editorial Council of the art and philosophy magazine *TechnoEtic Arts*, in Bristol, England, founded and directed by Roy Ascott.

Mr. Pimenta has been frequently invited, as professor and lecturer, by several institutions, among then the universities of New York, Georgetown, Lisbon, Florence, Lausanne, Tsukuba, São Paulo, Palermo, the Calouste Gulbenkian Foundation in Lisbon, the Monte Verita Foundation in Switzerland and the Technion Institute of Technology in Haifa, Israel.

He his founder and director of the Holotopia Academy: an informal institution oriented to music, art, philosophy and science, in the Amalfi Coast, Italy. He is also founder and director of the Foundation for Arts, Sciences and Technology – Observatory, in Trancoso, Portugal. In 2015 he was director of the World Art Forum, in Bolognano, Italy.

He lives between Locarno, Switzerland, which is his main residence, New York and Lisbon. His website is www.emanuelpimenta.net. In his website one can find his architectural projects, music projects, books, ebooks, photographic projects and essays, articles, and films.

some books by Emanuel Pimenta

Mr. Chico - A Zen Master in the Forests

John Cage - The Silence of the Music

Koellreutter - The Musical Revolutions of a Zen
Master

Medauar - O Homem que Sabia Demais

John Cage - Koan of Non Violence

John Cage: How to Change the World

One Hundred Years with John Cage

Virtual Notations (Music)

KAIROS - A Bird Orbitting Planet Earth (space architecture)

30 Years of Architecture

UIRA Orbital Olympic Village (space architecture)

Logical Traps

Art and Zen

Hidden Beings

Low Power Society

Mondo: Literature and Democracy

Walden Zero Project

Kirkos: A Dialogue Between Marcel Duchamp and
Josqin Des Près

Neapolis

Firenze: Mind Battle Fields of a Magical City

Ascona